Algebra 1 Ma
Handbo

Common Core Standards Edition

By:

Mary Ann Casey

B. S. Mathematics, M. S. Education

P. O. Box 328
Onsted, MI. 49265-0328

Acknowledgements

Gathering information and understanding about the implementation of the Common Core State Standards has been challenging. My friend and former colleague, Kimberly Knisell, Director of Math and Science in the Hyde Park (NY) School District, has been of great assistance in developing Algebra I Made Easy. Barbara Kandybowicz, who taught AP Statistics before she retired, was invaluable in helping me develop Unit 5. Proofreading math is a time consuming and frustrating experience – too many numbers and symbols besides the actual words! Thanks to Dorothy Draper-Silberg and Dick Swanson who did a painstaking proofread of the drafts of the units, and to Jennifer Criser-Eighmy who did a final review of everything but the math. Julieen Kane, graphic designer and assistant to my publisher, does a wonderful job putting the equations, graphs, and diagrams into publishing form despite claiming to "not be a math person." And lastly, thanks to Keith Williams, owner of the Topical Review Book Company and publisher of this handbook. Keith took a chance in 1996 that my classroom developed review material would make a good handbook. We have worked together successfully through several curriculum changes since then.

Introduction

__Algebra 1 Made Easy - Common Core Standards Edition__ is written to coordinate with the Common Core State Standards (CCSS). The CCSS is a nationally recognized set of standards that will help our math students become "college and career ready." This simply means they will be prepared to be successful in whatever they choose to do following graduation. As individual states adopt the CCSS, students and teachers across the country will be working hard to implement it. Instruction will be changed significantly in most classrooms leading to more experimentation and hands on learning. Students will be encouraged to "think outside the box" when analyzing problems and determining how to approach solving them. It will still be necessary for students to be able to use accepted mathematical processes to complete the analysis and solution. These methods are explained in this handbook. Algebra I Made Easy is designed to be a casual, student friendly handbook. It is a quick reference guide for students to use to obtain help with understanding which mathematical procedures and methods are acceptable and how to use them. This handbook contains the content covered by the CCSS, not the teaching techniques and learning styles that will be developed. When the student has analyzed the problem and chosen a desired method of approaching it, this handbook will help them "do the math" correctly. I wish everyone success with our implementation of the Common Core and hope this handbook helps you to reach that goal.

Sincerely,

MaryAnn Casey,
B.S. Mathematics, M.S. Education

ALGEBRA 1 MADE EASY
Common Core Standards Edition
Table of Contents

ALGEBRA 1 MADE EASY
Common Core Standards Edition
Table of Contents

FOUNDATIONS

- Algebraic Notation
- Real Numbers
- Numbers and Algebra
- Properties and Laws of Real Numbers and Polynomials
- Exponents
- Simplifying Radicals
- Pythagorean Theorem

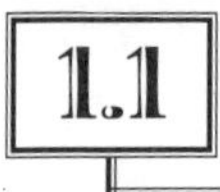

Algebraic Notation

Different notations are used to describe the elements or members of a set. Sometimes the elements are listed and in other cases a rule is used to define the members within the set.

DEFINITIONS

Universe or Universal Set: The set of all possible elements available to form subsets.

Example The set of real numbers.

Set: A group of specific items within a universe.

Example The set of integers.

Subset: A set whose elements are completely contained in a larger set.

Example The set of even integers is a subset of the set of integers.

Complement of a Set: Symbols are A', $\overline{A}$ or A^C. A' contains the elements of the universal set that are not in Set A.

Example If the universe is whole numbers from 2 to 10 inclusive, and Set $A = \{2, 4, 6, 8, 10\}$, then $A' = \{3, 5, 7, 9\}$.

Solution Set: All values of the variable(s) that satisfy an equation, inequality, system of equations, or system of inequalities.

Example In the set of real numbers, $\mathcal{R}$, the solution set for the equation $x^2 = 9$ is $\{3, -3\}$ because 3 and –3 are solutions to that particular equation. Write the solution set: $SS = \{-3, 3\}$

SYMBOLS

$\in$ means "is an element of". **Example** $100 \in$ set of perfect squares.

$\notin$ means "is not an element of". **Example** $3 \notin$ set of perfect squares.

$\varnothing$ or $\{\ \}$ are symbols for the "**empty set**" or the "**null set**". The empty set or null set has no elements in it.

Example If P is the set of negative numbers that are perfect squares of real numbers, then $P = \{\ \}$ or $P = \varnothing$.

Note: Do *not* use $\{\ \}$ and $\varnothing$ together. $\{\varnothing\}$ means the set containing the element $\varnothing$. It does not mean the empty set or null set.

Notation: Ways to write the elements of a set.

1.1

Roster Form: A list of the elements in a set (Set A).

Example $A = \{2, 4, 6, 8, 10\}$

Set Builder Notation: A descriptive way to indicate the elements of a set.

Example The set of real numbers between 0 and 10, inclusive, would be written: $\{x : x \in \mathcal{R}, 0 \leq x \leq 10\}$ or $\{x \mid x \in \mathcal{R}, 0 \leq x \leq 10\}$. The colon or the line both mean "such that." This is read, "The set of all values of x such that x is a real number and 0 is less than or equal to x and x is less than or equal to 10."

Inequality Notation: Inequality symbols are used to define the elements in the set. Symbols for exclusive are $<$ *and* $>$. Symbols for inclusive are $\leq$ *and* $\geq$.

Examples

❶ $-5 \leq x < 7$ means all values of x that are equal to or greater than -5 and less than 7. This includes -5 and excludes 7.

❷ $x < 3$ or $x \geq 10$ means the elements of the set that are less than 3 or the elements that are greater than or equal to 10. This excludes 3 and includes 10.

Interval Notation: Use (,) or [,] to indicate the end elements in a list of elements. Parentheses show the number is not included; brackets show that it is.

Example Set $E = \{x: 3 \leq x < 6\}$ would be written $[3, 6)$.
Set $F = \{x: 5 \leq x \leq 8\}$ would be written $[5, 8]$.
Set $G = \{x: 0 < x < 5\}$ would be written $(0, 5)$.

Examples

Elements defined in words	All values of x that are greater than or equal to 2 and less than 6.
Set Builder Notation	$\{x \mid 2 \leq x < 6\}$
Inequality Notation	$2 \leq x < 6$
Interval Notation	$[2, 6)$

Two common symbols for sets of numbers:

$\mathcal{R}$ – the set of real numbers. $\mathcal{R} = \{-\infty, \infty\}$ The "reals" contain all the numbers from negative infinity to positive infinity.

$\mathcal{W}$ – the set of whole numbers. $\mathcal{W} = \{0, 1, 2, 3, \ldots\}$ The whole numbers are 0, 1, 2, 3, and so on. The three dots indicate "and so on" following the pattern established by at least 3 previous elements.

- The 3 dots can also be used on the negative side of a pattern.

Example Integers $< 0 = \{\ldots, -3, -2, -1\}$

1.2 Real Numbers

Sets of Numbers:

All the sets below are part of the Set of Real Numbers, $\mathcal{R}$. The REAL NUMBERS consist of *all* the numbers on a number line. If it were possible to place a point on the number line for each real number, the line would be completely filled in and extend forever (infinitely) in the negative and in the positive directions. (These are also referred to as negative infinity and positive infinity.) Real Numbers cannot be listed as a set. Write SS = {Real Numbers} if needed. The Real Numbers are composed of two major sets of numbers called RATIONAL numbers and IRRATIONAL numbers. Each of these sets contains several subsets.

The Set of Real Numbers includes:

- ***Positive Numbers*** = $\{x \mid x > 0\}$.
  This is read, "The set of positive numbers contains all the values of x such that x is greater than zero." In other words, all the numbers in this set are greater than zero.

- ***Negative Numbers*** = $\{x \mid x < 0\}$.
  Read as: All the values of x less than zero

- ***Zero*** = $\{x \mid x = 0\}$.

1. Rational Numbers: Symbol is (Q). Rational numbers are defined as the ratio of two integers and can be written in the form $\frac{a}{b}$ where a and b are integers and $b \neq 0$ (Division by zero is undefined). It is acceptable for b to be 1, so a rational number like 3 could be written $\frac{3}{1}$. Rational numbers include these sub-sets.

a) **Whole Numbers** or $\mathcal{W} = \{0, 1, 2, 3, ...\}$.
This is read, "The set of whole numbers contains the numbers 0, 1, 2, 3, and so on." The "and so on" means that the pattern of numbers shown is to be continued.

b) **Integers:** $Z = \{..., -2, -1, 0, 1, 2, ...\}$ *Subsets of integers include*:

- ***Odd Integers*:** Positive or negative. Have a remainder when divided by 2. Odd Integers = $\{..., -5, -3, -1, 1, 3, 5, ...\}$
- ***Even Integers*:** Positive or negative. Divisible by 2. Include 0. Even Integers = $\{..., -6, -4, -2, 0, 2, 4, 6, ...\}$
  ****Zero is an even integer, but it has no sign.****
- ***Positive Integers*:** $\{1, 2, 3, ...\}$ [does *not* include 0]
- ***Non-Negative Integers*** = $\{\mathbf{0}, 1, 2, 3, ...\}$ [includes Positives & 0]
- ***Negative Integers*:** $\{..., -3, -2, -1\}$ [does *not* include 0]
- ***Non-Positive Integers*:** $\{..., -3, -2, -1, \mathbf{0}\}$ [includes Negatives & 0]
- ***Natural Numbers or Counting Numbers*:** $\mathcal{N} = \{1, 2, 3, ...\}$

c) **Fractions:** A ratio of two integers with a denominator that is not equal to zero, like $\frac{2}{3}$ *or* $\frac{5}{4}$.

d) **Terminating Decimals:** Such as 0.24 which can be written $\frac{24}{100}$

e) **Repeating Decimals:** Such as $0.\overline{37}$

Example Explain how to change a repeating decimal into a rational number in the form $\frac{a}{b}$, $b \neq 0$.

Solution: Make an equation with n = repeating decimal. Multiply both sides of the equation so the repeating part of the decimal begins right after the decimal point. Then subtract the original equation from the new one. When the subtraction is completed, only zeroes will appear to the right of the decimal. Solve for n and simplify if necessary.

1. Using $n = 0.666\ldots$ as the repeating decimal number, multiply both n and the number by 10.

$$\begin{aligned} 10n &= 6.66\overline{6} \\ -\ n &= 0.66\overline{6} \\ \hline 9n &= 6.000 \\ n &= \frac{6}{9} \text{ or } \frac{2}{3} \end{aligned}$$

> 1. Multiply both sides of the equation by 10.
> 2. Subtract the first equation from the 2nd.
> 3. Solve for n.
> 4. Simplify if necessary.

2. If $n = 0.125125\ldots$, then it is necessary to multiply both sides of the equation by 1000 in order to line the decimals up correctly for subtraction. (See solution 1.)

$$\begin{aligned} 1000n &= 125.125\overline{125} \\ -\ n &= 0.125\overline{125} \\ \hline 999n &= 125 \\ n &= \frac{125}{999} \end{aligned}$$

f) **Roots of "perfect" numbers** which are rational.

The square root of 25; $\sqrt{25} = 5$.

The cube root of –8; $\sqrt[3]{-8} = -2$.

$\sqrt{0.04} = 0.2$ *or* $\sqrt{225} = 15$.

***Note*:** There is an infinite number of rational numbers and they cannot be written as a listed set. Write SS = {rational numbers} if needed.

2. **Irrational Numbers:** Real numbers which cannot be expressed as the ratio of two integers.

   a*) *Non-repeating, non-terminating decimals like 0.354278...

   b*)** The ***roots of numbers which do not calculate to a rational number.

   The square root of 12 which ≈ 3.4641016 (rounded).
   $\sqrt{2}$ is irrational as is $\sqrt{5}$.
   $\sqrt[3]{16}$ is irrational also.

   ***c*)** **Pi *or* π** is also irrational. The "π" *button on the calculator must be used to compute formulas that contain* π.
   Rounding to 3.14 is *not* permitted.

   ***Note*:** When a calculator is used to work with irrational numbers, rounding is not permitted. To "show all work" write down on the paper all the digits in the calculator display. Leave the entire number in the calculator until the end of the problem. When directions say to round, it should be the final step of a problem. The approximate value of π, such as 3.1416, 3.14, or 22/7, are *unacceptable* unless otherwise specified.

The diagram below represents the relationship between subsets of real numbers. The universal set, shown by the rectangular box, is the set of Real Numbers. Subsets are shown as circles. Subsets of the Reals are Irrational Numbers and Rational Numbers. Subsets of Rationals are Integers and its subsets are Whole Numbers and Counting Numbers.

Real Numbers

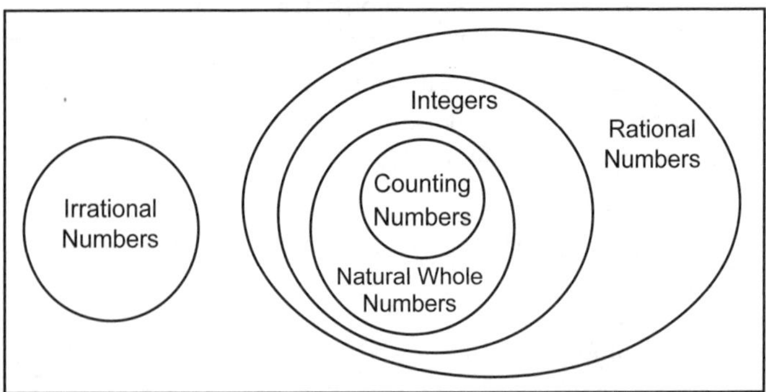

1.3 Numbers and Algebra

Number Line: A representation of all the real numbers in our number system. Each point on a number line corresponds with a REAL NUMBER. Numbers are shown on the number line with zero in the center, the negative numbers on the left side and the positive numbers on the right side of zero. The line extends infinitely in both directions. We often use only a part of the number line — just showing the numbers that are relevant to our current work. Equally spaced tick marks show the placement of the whole numbers and are labeled. All fractions, decimals, square roots, both positive and negative, are located in relation to the labels shown for the whole numbers.

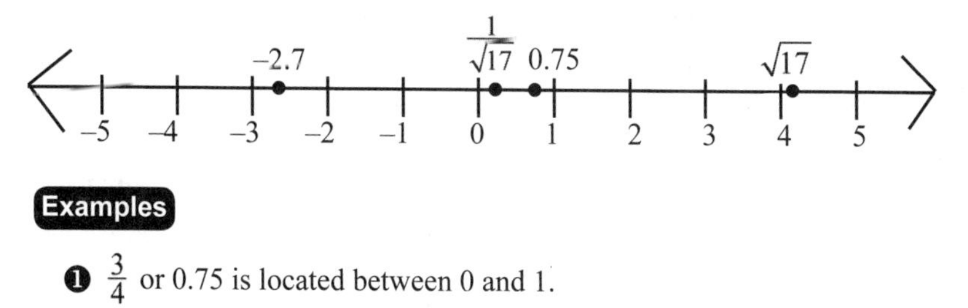

Examples

❶ $\frac{3}{4}$ or 0.75 is located between 0 and 1.

❷ –2.7 is located between –2 and –3.

❸ $\sqrt{17} \approx 4.123$ which is between 4 and 5.

❹ $\frac{1}{\sqrt{17}} \approx 0.243$ 0.243 is between 0 and 1, quite close to zero.

Order: On a number line, the numbers get larger as we read toward the right. In the example above, –2.7, the number furthest to the left, is the smallest (least) number used. $\sqrt{17}$ is the largest number in our example and its point is furthest to the right on the number line. Using < and > signs, we would say: $-2.7 < \frac{1}{\sqrt{17}} < 0.75 < \sqrt{17}$. If you are uncertain where a number is located on a number line, convert the number to a decimal and it will be easier to find on the number line. (Use at least 3 places after the decimal point when changing into decimals.) Then the numbers can be easily compared by examining their relative positions.

1.3

Reciprocal: The product of a number and its reciprocal is 1. The reciprocal of a number can be formed by making the number into a fraction and then inverting the fraction (flipping it upside down).
(See also negative exponents page 17)

Examples *2 and* $\frac{1}{2}$*, or* $\frac{2}{3}$ *and* $\frac{3}{2}$*, or –5 and* $-\frac{1}{5}$*, or x and* $\frac{1}{x}$

***Note*:** If variables (letters) are used instead of numbers in a general question about the order of numbers, choose a number (like 5) and substitute it into the problem. You can see how that particular number works and then you can usually apply that information to the problem in general.

Examples For the set of positive real numbers, is $1 \leq x \leq \frac{1}{\sqrt{x}}$ true or false? Substitute a number for x that makes sense here, like 9. At first it seems true since $1 \leq 9$ --- but 9 is not $\leq \frac{1}{3}$ so that makes the expression false if x is 9. One counter-example is enough to make it false, but check once more. Substitute a small number, like 0.04, and see if it is still false. Since 1 is not ≤ 0.04, this example is already false. Trying a couple of different numbers and putting them on the number line is helpful if you are "stuck".

Absolute Value **(| |):** Absolute value is the distance from the number inside the symbol (| |) to zero on a number line.

Rules To Remember:

- It is always positive.
- Do the inside work first, then make that answer positive.
- The absolute value of 0 is zero.
- A negative sign to the left of the absolute value sign means to find the absolute value of the "inside" work and then make that answer negative. (See Example 3)
- A number in front of the absolute value sign means to find the absolute value inside the symbols and then multiply by the outside number. (See Example 4)

Examples

❶ $|5| = 5$

❷ $|-8| = 8$

❸ $-|3+4| = -|7| = -7$

❹ $2|2-15| = 2|-13| = 2(13) = 26$

1.3

OPERATIONS WITH SIGNED NUMBERS

Work left to right using the correct order of operations. If there are more than two numbers to work on, do them in pairs -- always using the order of operations. (See page 10)

Addition: LIKE signs (all + or all –; all the same): Keep the sign, add the absolute values.

Examples

❶ $12 + 5 = 17$

❷ $(-5) + (-3) = -8$

❸ $(-4) + (-2) + (-7) = (-6) + (-7) = -13$

UNLIKE signs (one +, one –; mixed or different signs) Find the difference between the absolute values of the numbers. Keep the sign of the number with the larger absolute value.

Examples

❶ $12 + (-8) = 4$

❷ $20 + (-30) = -10$

❸ $(-8) + 5 + (-9) = (-3) + (-9) = -12$

Subtraction: Change subtraction sign to an addition sign, and change the sign of the number after the subtraction sign. Now follow the addition rules.

Examples

❶ $(-9) - (+6) = -9 + (-6) = -15$

❷ $12 - 8 + 15 = 12 + (-8) + 15 = 4 + 15 = 19$

❸ $(-10) - (-7) - 6 = (-10) + 7 + (-6) = -9$

1.3

Multiplication and Division (Two Numbers):

Like Signs: (two negatives or two positives). Answer is positive.

Examples

❶ $(3)(4) = 12$

❷ $(-5)(-4) = 20$

❸ $12 \div 4 = 3$

Unlike Signs: (one of each). Answer is negative.

Examples

❶ $(-2)(3) = -6$

❷ $15 \div (-5) = -3$

Division in Fractions: The fraction line acts like a grouping symbol. Using the correct order of operations, do all the work in the numerator (top) of the fraction. Do all the work in the denominator (bottom) of the fraction using the correct order of operations. Then divide the numerator by the denominator. Follow the rules for signs as needed in each part of the problem.

Example $\frac{(14-2)-(3+3)}{-12} = \frac{12-6}{-12} = \frac{6}{-12} = -\frac{1}{2}$ *or* -0.5

Division by Zero is undefined: If a fraction has zero as a denominator, the division cannot be performed. The fraction is said to be "undefined."

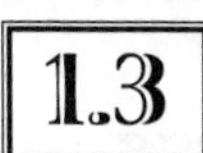

ORDER OF OPERATIONS

When performing operations on numbers (or on algebraic terms), it is necessary to follow these rules that tell which part of the work to do first, second, etc. Read the problem from left to right, just as you read words, and follow these steps:

Steps:

1) **Parenthesis** — Do whatever work you can inside a parenthesis to simplify the problem. The fraction line in a fraction acts like a parenthesis -- do the work in the numerator of the fraction and the work in the denominator of the fraction before continuing.

2) **Exponents** — Apply any exponents in the problem to their bases. This includes a parenthesis with an exponent or a term with an exponent raised to another power.

3) **Multiply and Divide** — Do these in order as they appear as you read left to right. This includes using the distributive property.

4) **Add and Subtract** — Do these in order as they appear reading left to right.

Note: Remember to always work left to right -- within a parenthesis or in a fraction as well as across the entire problem.

The example below shows a combination problem that requires several steps to simplify. (See also Monomials and Polynomials page 45)

Example $3(x + 3x - 1)^2 - (8x^2 + 4)$

Steps:

1) Do work inside the ( ) first.	$3(4x - 1)^2 - (8x^2 + 4)$
2) Exponent applied to (4x – 1) only.	$3(4x - 1)(4x - 1) - (8x^2 + 4)$
3) Use F. O. I. L.	$3(16x^2 - 4x - 4x + 1) - (8x^2 + 4)$
4) Collect like terms inside the ( ).	$3(16x^2 - 8x + 1) - (8x^2 + 4)$
5) Use the Distributive Property to multiply and remove ( ).	$48x^2 - 24x + 3 - 8x^2 - 4$
6) Collect like terms to simplify.	$40x^2 - 24x - 1$

1.4

Properties and Laws of Real Numbers & Polynomials

Why do we care about the properties? These fundamental laws allow us to manipulate the terms in an equation. We can move things around, transfer parts of an equation to the opposite side of the equal sign, and use the identities to help isolate the variable. Closure applies to polynomials. The addition of two polynomials results in a sum that is a polynomial. The associative property is also applicable to polynomials under addition. This means the grouping can be changed. The additive identity of zero exists and each polynomial has an additive inverse. The commutative property is also true for the addition of polynomials. Combining these properties and identities gives us the flexibility we need to solve problems and equations involving polynomials algebraically These laws and properties apply to real numbers and polynomials.

Commutative Property of Addition and Multiplication: The POSITION of the numbers can be changed without changing the answer.

Addition: $a + b = b + a$

Example $5 + 1 = 1 + 5$
$x + 3 = 3 + x$

Multiplication: $ab = ba$

Example $(-5)(-2) = (-2)(-5)$
$5(x) = x(5)$

Note: The commutative property is ***not*** true with respect to subtraction or division of real numbers or polynomials.

Subtraction: $5 - 3 \neq 3 - 5$
$x - 4 \neq 4 - x$

Division: $\frac{2}{3} \neq \frac{3}{2}$
$(x + 1) \div (x + 2) \neq (x + 2) \div (x + 1)$

Associative Property of Addition and Multiplication: The GROUPING of the numbers can be changed without changing the answer.

Addition: $(a + b) + c = a + (b + c)$

Example $(5 + 4) + 2 = 5 + (4 + 2)$
$(x + 2) + y = x + (2 + y)$

Multiplication: $(ab)\,c = a(bc)$

Example $(3 \cdot 2)(4) = 3(2 \cdot 4)$
$(3 \cdot x)(7) = 3(x \cdot 7)$

Note: The associative property is ***not*** true for subtraction or division of real numbers.

Subtraction: $(3 - 5) - 12 \neq 3 - (5 - 12)$
$(x - 1) - (x + 2) \neq x - (1 - x) + 2$

Division: $(12 \div 3) \div 4 \neq 12 \div (3 \div 4)$
$(2x + 1) \div (x + 2) \neq 2x + (1 \div x) + 2$

1.4

Distributive Property of Multiplication Over Addition: Multiply each term on the inside by the number or letter on the outside of the parenthesis.

Examples

❶ $a(b + c) = ab + ac$

❷ $2(3 + 5) = 2(3) + 2(5)$

❸ $-3(x + 2) = -3x - 6$

❹ $-5(2x + 5 + 2x) = -5(4x + 5) \Rightarrow$ then $-20x - 25$ is the final answer.

❺ $2x(3x - 4) = 6x^2 - 8x$

Use the positive or negative sign before the outside number with the number when you multiply. This removes the parenthesis and takes care of the needed sign changes. Combine terms inside the parenthesis before you use the distributive property if possible. If the number on the outside is a fraction, be sure to multiply each term inside the parenthesis by the entire fraction.

***Special Note:* Use the Distributive Property for Subtraction Problems**

A parenthesis with subtraction sign in front of it can be handled using the Distributive Property. Put "1" outside the parenthesis between "–" and the ( ). Use the Distributive Property by multiplying by –1. This will automatically take care of the sign changes needed for subtraction and remove the ( ), then simplify (Combine like terms).

Examples

❶ $3x - 4 - (7 + x) \Rightarrow 3x - 4 - 1(7 + x) \Rightarrow 3x - 4 - 7 - x \Rightarrow 2x - 11$

❷ If a problem says "take $(3x - 4)$ from $(12x + 5)$", then do this:

Steps:

1) Set-up	$(12x + 5) - (3x - 4)$	"From" number is first in set-up
2) Get ready	$(12x + 5) - 1(3x - 4)$	Put a "1" between – and $(3x - 4)$.
3) Remove ( )	$12x + 5 - 3x + 4$	Use Distributive Property and multiply by –1.
4) Simplify	$9x + 9$	Collect Like Terms.

❸ If the directions say "SUBTRACT" and the problem is already set up:

Step 1) is already done.	**Step 2)**	**Step 3), then 4)**
$(5x^2 - 12x + 14)$	$(5x^2 - 12x + 14)$	$5x^2 - 12x + 14$
$-(x^2 + 13x - 25)$	$-1(x^2 + 13x - 25)$	$+(-x^2 - 13x + 25)$
		$4x^2 - 25x + 39$

Note: Sometimes the directions say "subtract these problems" and there may not be a subtraction sign in front of the lower expression. Put in your own ( ) and –1 if needed, then go on. READ the directions.

1.4

Identity Element: An element that can be used with the given operation on any member of the set without changing its value is called the identity element for that operation. If * is a binary operation on set **S**, then "e" is an identity element if, for all a in **S**, $a * e = e * a = a$.

Identities – Real Numbers and Polynomials

Additive Identity: Zero (0) is the additive identity. A number or polynomial is unchanged by adding zero to it or by adding the number to zero.

Examples ❶ $5 + 0 = 0 + 5 = 5$ ❷ $(2x + 1) + 0 = 0 + (2x + 1) = 2x + 1$

Multiplicative Identity: One (1) is the identity element for multiplication. A number or polynomial multiplied by one is unchanged.

Examples ❶ $(5)(1) = (1)(5) = 5$ ❷ $(x + 4)(1) = 1(x + 4) = x + 4$

Inverse Element: In a system with an identity element "e" there is an inverse "b" such that $a * b = b * a = e$. When the given operation is performed on an element and its inverse, the result is the identity element.

Inverses – Real Numbers and Polynomials

Additive Inverse: The additive inverse of a number in the real number system is the opposite number. (The same number but with the opposite sign.) Since 0 is the identity element for addition in the real numbers, then –4 is the additive inverse of 4. Polynomials have the same property.

Examples ❶ $(-4) + (4) = 0$
Since $(-3.2) + (3.2) = 0$
and $(3.2) + (-3.2) = 0$,
3.2 is the additive inverse of –3.2.

❷ $(x - 5) + [-1(x + 5)] = 0$

Multiplicative Inverse: The reciprocal of a real number is its multiplicative inverse. The product of a number and its reciprocal is 1 which is the identity element for multiplication of real numbers.

Examples ❶ $-3\left(-\frac{1}{3}\right)=1$ and $\left(\frac{2}{3}\right)\left(\frac{3}{2}\right)=1$ ❷ $(x + 2) \bullet \frac{1}{(x + 2)} = 1$

The multiplicative inverse of a negative number is still its reciprocal — be sure and keep the reciprocal as a negative number.

Note: The reciprocal of 1 is 1 and the reciprocal of –1 is –1. These properties, law, and definitions can be applied to other systems of mathematics that may include letters and operations that with which we are not yet familiar.

Multiplication Property of Zero: The product of a number or polynomial multiplied by zero is zero.

Examples ❶ $5 \bullet 0 = 0$ ❷ $(x^2 + 12x + 7)(0) = 0$

Binary Operation: A rule that is applied to two elements of a set. Multiplication, addition, subtraction and division are binary operations.

Closure: A set, **S**, is closed under a binary operation (*) if for all "a" and "b" in **S**, $a * b$ equals an element of set **S**. In other words, if a binary operation performed on any two members of a set always results in a member of the set, then the set has closure with respect to that operation. To determine if a set has closure with respect to an operation, perform the operation using all members of the set. If the answer is always a member of the set, it has closure with respect to the operation.

- The set of **Whole Numbers,** $\mathcal{W}$, is closed with respect to (or under) addition and multiplication; but not under subtraction and division.

Examples

❶ **Addition:** $2 + 5 = 7$. 2, 5, and 7, are all elements of $\mathcal{W}$. There are no examples of addition of whole numbers which would not result in a whole number answer.

❷ **Multiplication:** $2 \cdot 6 = 12$. 2, 6, and 12 are all whole numbers. The product of any two whole numbers is a whole number.

❸ **Subtraction and Division:** The quotient or the difference resulting from division or subtraction of two whole numbers may not be a whole number. The set is not closed under subtraction or division. $2 - 7 = -5$; $7 \div 2 = 3\frac{1}{2}$. Negative 5 and $3\frac{1}{2}$ are not whole numbers. *

- The set of **Polynomials** has closure under addition, subtraction, and multiplication but is not closed with respect to division.

Examples

❶ **Addition:** $(x^2 + 7x + 2) + (x + 10) = x^2 + 8x + 12$
Polynomial sum shows closure.

❷ **Subtraction:** $(3x^3 + 12) - (2x^3 - 2x + 1) = x^3 + 2x + 11$
Polynomial difference shows closure.

❸ **Multiplication:** $(x^2 + 1)(x - 3) = x^3 - 3x^2 + x - 3$
Polynomial product shows closure.

❹ **Division:** $(x^2 + 1) \div (x - 5) = \frac{x^2 + 1}{x - 5}$: Quotient is ***not*** a polynomial showing the set is not closed under division.*

* These are called **counterexamples** as they each demonstrate at least one possible result of the operation that is not a member of the original set. A counterexample is useful to prove something is not always true.

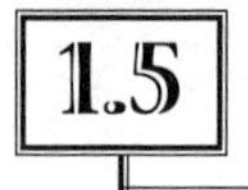

Exponents

Exponent: It indicates how many times to use its base as a factor to make a product. An exponent is used only with whatever number or term is directly to its left. If the exponent is outside a parenthesis, then use the exponent with everything in the parenthesis. If the exponent is next to a term instead of a parenthesis, it goes only with its immediate neighbor to the left. When simplified, a problem with exponents should have an answer where only the variables still have exponents. Numbers with exponents should be multiplied out. (See also Rational Exponents, page 29)

Examples

❶ $(3xy)^3 = (3xy)(3xy)(3xy) = 27x^3y^3$

❷ $(-4)^2 = (-4)(-4) = 16$

❸ $-4^2 = -(4)(4) = -16$

(Check this one out! The exponent goes with the $(2x)$ and not with the -3)

❹ $-3(2x)^3 = -3(2x)(2x)(2x) = -3(8x^3) = -24x^3$

Multiplication: When the bases are alike, ADD the EXPONENTS. MULTIPLY the COEFFICIENTS.

Examples

❶ $(4xy^3)(5xy^2) = 20\,x^2y^5$

❷ $(2x^4)(-3x^{-2}) = -6x^2$

Division: When the bases are alike, SUBTRACT the EXPONENTS. DIVIDE the COEFFICIENTS.

Example

$$\frac{6x^3y^4}{2x^2y} = 3x^{3-2}y^{4-1} = 3xy^3$$

Powers: When a term with an exponent is used as a base with another exponent, it is called "raising a power to a power." Coefficients are treated as if they have an exponent of 1. Multiply each exponent inside the parenthesis by the exponent on the outside. Be sure to include the "invisible" exponent 1 next to the coefficient. Then simplify if possible.

Examples

❶ $(x^5)^4 = x^{20}$

❷ $(x^2y)^3 = x^6y^3$

❸ $(3x^4y^2)^3 = (3^1x^4y^2)^3 = 3^3x^{12}y^6 = 27x^{12}y^6$

Zero as an exponent: ANY base (except 0) with an exponent of 0 is equal to **1**. Replace the indicated base with 1 in the problem before continuing.

Note: A base of zero raised to the zero power $(0)^0$ is undefined.

Examples

1. $x^0 = \mathbf{1}$
2. $x^2y^0 = x^2 \bullet \mathbf{1} = x^2$
3. $4m^0 = 4(\mathbf{1}) = 4$
4. $\left(42(3250)^3(x^4)\right)^0 = \mathbf{1}$

Negative Exponents: Make a fraction putting any base with a negative exponent in the denominator. This will make the exponent positive. The coefficient of the base and any variable with positive exponents are in the numerator. Negative exponents are undefined if the base = 0. Remember the base of the exponent can be contained in a ( ).

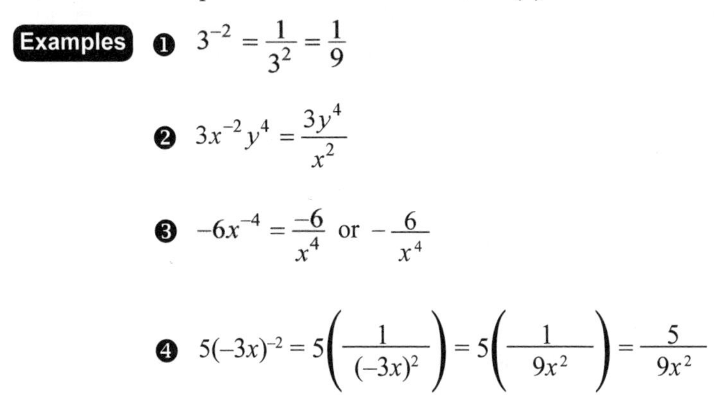

Examples

1. $3^{-2} = \frac{1}{3^2} = \frac{1}{9}$
2. $3x^{-2}y^4 = \frac{3y^4}{x^2}$
3. $-6x^{-4} = \frac{-6}{x^4}$ or $-\frac{6}{x^4}$
4. $5(-3x)^{-2} = 5\left(\frac{1}{(-3x)^2}\right) = 5\left(\frac{1}{9x^2}\right) = \frac{5}{9x^2}$

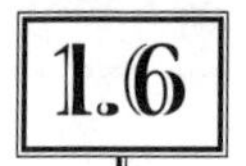

SIMPLIFYING RADICALS

Terminology and Definitions:

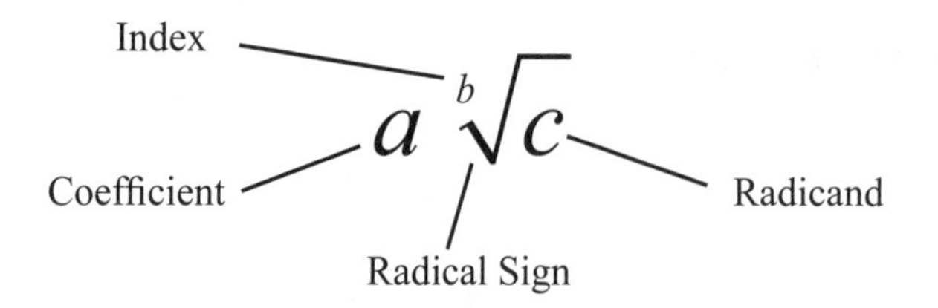

A **Radical Sign** ($\sqrt{\ }$) in a problem indicates that a root of the number under the "radical" is to be found. If no index is given, then the positive square root of the radicand is the answer.

Example $\sqrt{81} = 9$

We know that the square root of 81 could also be (– 9), but the radical sign indicates use of the positive root only. If both square roots of a number are to be used, a ± sign is in front of the radical.

Example $\pm\sqrt{81} = \pm 9$ or $+9$ and -9

When solving a quadratic equation (see page 95) we have to indicate that the positive and negative roots are to be used.

Example *If* $x^2 = 25$, then $x = \pm\sqrt{25}$, and $x = \pm 5$

Note: Although all positive numbers have both a positive and a negative square root, we will refer to positive square roots in this section unless directed otherwise. When finding a square root, if the radicand is a negative number, there is no real square root.

Root: A root is a factor of the radicand which is multiplied by itself a given number of times to produce the number under the $\sqrt{\ }$. A square root means to multiply the answer by itself to get the radicand. A cube root, when multiplied by itself three times has a product equal to the radicand.

Example The square root of 81 is 9 because $9 \bullet 9 = 81$

Radicand: The number (and/or variables) under the radical sign.

1.6

Index: The small number over the left edge of the radical sign which indicates the type of root to be found. If there is no number there, it is assumed to be "2".

❶ $\sqrt[3]{8} = 2$. 3 is the index. Find the cube root of 8 which is 2.

❷ $\sqrt{16}$ There is no index so an index of 2 is understood. Find the square root of 16 which is 4.

Coefficient: Refers to the number outside the radical sign. Since radical signs are often treated as variables, "coefficient" seems appropriate. Multiply the root by the coefficient for a final answer.

Example $7\sqrt[3]{64}$ 7 is the "coefficient".
3 is the index.
64 is the radicand.

In words: Find the cube root of 64, then multiply that number by 7. The cube root of 64 is 4. The final answer is $7(4) = 28$.

"Like" Radicals: Have the same index and the same radicand.

Example $\sqrt{5}$ *and* $2\sqrt{5}$ are "like." $\sqrt{2}$ and $\sqrt{3}$ are not "like radicals."

EXACT answers require that a root that is irrational be left in radical form. If an estimated answer is needed, use the $\sqrt{\ }$ button on your calculator and then round the answer to the place indicated in the problem.

Rational and Irrational Roots:

1. The square root of a positive number that is a perfect square is a rational number. No $\sqrt{\ }$ will be visible in the answer.

    $\sqrt{36} = 6$

2. The square root of any positive number that is not a perfect square is an irrational number. Irrational square roots are left "in simplified radical form" unless directed otherwise and a radical sign will be part of the answer.

   Examples

   ❶ $\sqrt{8} = 2\sqrt{2}$ in simplest radical form. (This is an EXACT answer.)

   ❷ $\sqrt{8} \approx 2.8284271$ Write the full display of the calculator on your paper unless directed to round.

   ❸ If $x^2 = 2$; $x = \pm\sqrt{2}$. x is irrational. When $x^2 = 5$, $x = \pm\sqrt{5}$ and x is irrational.

Simplifying Radicals: First test the number under the radical to see if it is a perfect square. Use the $\sqrt{}$ button on your calculator - if the answer is a whole number, you are done! If not - follow this procedure to simplify:

Steps

1) Break the number under the radical sign (the radicand) into its prime factors.
2) Remove from the radical any pair of numbers that are factors - placing one of each pair on the outside of the radical sign. The "other one" of the pair of numbers disappears because by removing the pair from the radical, you have actually found the square root of that pair of factors.
3) Continue this process until as many pairs of factors - or "perfect squares" as possible are removed from under the $\sqrt{}$.
4) The outside factors are multiplied together to make the "coefficient" of the radical.
5) Multiply back together any factors remaining under the radical sign.

Example $\sqrt{540} = \sqrt{3 \bullet 3 \bullet 3 \bullet 2 \bullet 2 \bullet 5} = 3 \bullet 2\sqrt{3 \bullet 5} = 6\sqrt{15}$

Simplifying With Variables: Use the same process shown above, but include the variables in the factoring process.

Example $\sqrt{12x^3y^2} = \sqrt{2 \bullet 2 \bullet 3 \bullet x \bullet x \bullet x \bullet y \bullet y} = 2 \bullet x \bullet y\sqrt{3 \bullet x} = 2xy\sqrt{3x}$

This example has a pair of 2's, a pair of x's and a pair of y's. Find the square root of each pair and move it to the outside of the radical sign. Multiply the outside factors together. Then multiply the inside factors together.

Note: As you practice simplifying radicals, the perfect squares or cubes will become familiar.

Adding or Subtracting Radicals:

Steps

1) Only "like" radicals can be added. Sometimes it is possible to make them "alike" by simplifying the original problem. Both examples below are shown being simplified first.
2) Add (or subtract) the coefficients.
3) Keep the radicand unchanged.

Examples

❶ $\sqrt{8}+\sqrt{32}=2\sqrt{2}+4\sqrt{2}=6\sqrt{2}$

❷ $3\sqrt{50x}-85\sqrt{2x}=15\sqrt{2x}-85\sqrt{2x}=-70\sqrt{2x}$

4) Some radicals cannot be added or subtracted. Simplify first, as usual. When it becomes evident that they cannot be simplified to match, leave the answer as shown here. The answers will have several simplified parts.

Examples

❶ $\sqrt{15}+\sqrt{20}=\sqrt{15}+2\sqrt{5}$ This one has two radicals for the answer.

❷ $\sqrt{25}-\sqrt{60}=5-2\sqrt{15}$ This one has a whole number and a radical for the answer.

Multiplying or Dividing Radicals:

Steps

1) The radicands do not have to match.
2) Multiply or divide the coefficient of one by the coefficient of the other.
3) Multiply or divide the radicands. Notice that when working on division problems both radicands are put into one radical sign (See example #3 below).
4) Simplify if possible.

Examples

❶ $\sqrt{5}\bullet 6\sqrt{10}=6\sqrt{5\bullet 10}=6\sqrt{50}=30\sqrt{2}$

❷ $x\sqrt{3y}\bullet 4x\sqrt{5z}=4x^2\sqrt{15yz}$

❸ $\frac{4\sqrt{12}}{2\sqrt{6}}=\frac{4}{2}\sqrt{\frac{12}{6}}=2\sqrt{2}$

❹ $\frac{24\sqrt{6}}{3}=\frac{24}{3}\sqrt{6}=8\sqrt{6}$

1.7 Pythagorean Theorem

Use The Pythagorean Theorem for: finding the length of the hypotenuse of a right triangle, finding the length of a diagonal of a rectangle or square (if you know the sides), finding either of the two legs if you know one of them and the hypotenuse, finding both legs if it is an isosceles right triangle and you know the hypotenuse. The equation for the Pythagorean Theorem is $c^2 = a^2 + b^2$.

Example If the hypotenuse of a right triangle is 12 and one leg is 7, find the length of the other leg to the nearest hundredth.

$a = 7 \quad c = 12 \quad b = ?$

$c^2 = a^2 + b^2$

$(12)^2 = 7^2 + b^2$

$144 = 49 + b^2$

$-49 \quad -49$

$95 = b^2$

Steps

1) Use your calculator. $\quad b = \sqrt{95}$
2) Write the answer to at least one place past the rounding place. $\quad b = 9.746...$
3) Then round as indicated. $\quad b = 9.75$

Example Jerry is building a set of stairs that have a rise of 108 inches in height. The run is 144 inches in length. What length piece of lumber does he need to cut in order to make the stringer? (The stringer is the piece of lumber that supports the steps as shown in the diagram.)

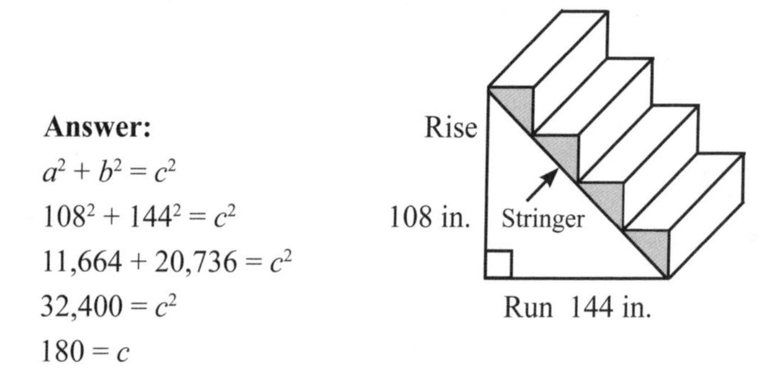

Answer:

$a^2 + b^2 = c^2$

$108^2 + 144^2 = c^2$

$11{,}664 + 20{,}736 = c^2$

$32{,}400 = c^2$

$180 = c$

180 inches or 15 feet is the length of lumber that Jerry needs to cut to make the stringer.

Example Macie has a broken leg and is in a cast and uses crutches. She wants to be able to prepare meals without using the crutches. The doctor wants to know how far she would have to walk between her appliances. Her kitchen is sketched on the grid and shows the placement of the sink, stove, and refrigerator. She works at the middle of each appliance. To the nearest tenth, what is the distance Macie would have to walk between the sink and refrigerator?

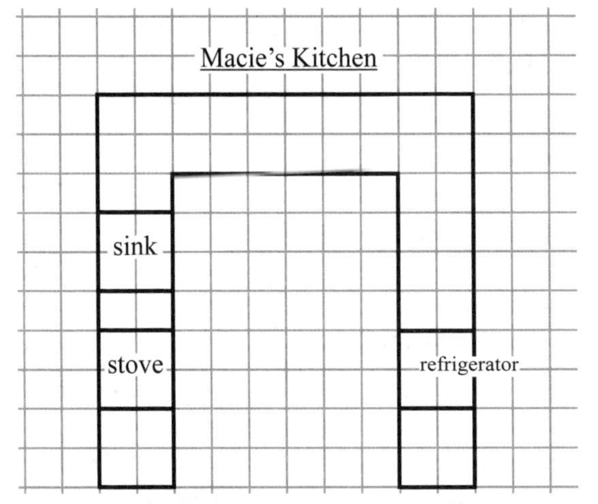

Solution: In order to find the distance from the sink to the refrigerator, make a right triangle on the grid using the distance from the stove to the refrigerator, 6 feet, as one leg of a right triangle. The other leg is the distance from the stove to the sink, 3 feet. The two legs form a right angle. The distance from the sink to the refrigerator is the hypotenuse of the right triangle. Use the Pythagorean Theorem.

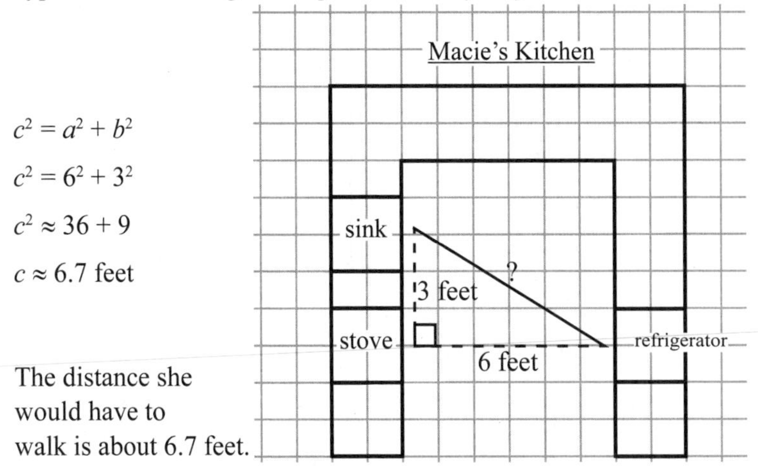

$c^2 = a^2 + b^2$

$c^2 = 6^2 + 3^2$

$c^2 \approx 36 + 9$

$c \approx 6.7$ feet

The distance she would have to walk is about 6.7 feet.

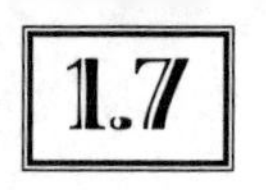

Converse of the Pythagorean Theorem says that if the sum of the square of the lengths of the legs equals the square of the hypotenuse, then the triangle is in fact a right triangle. So if $a^2 + b^2 = c^2$, then the triangle is a right triangle.

Example A triangle has side lengths of 12, 8, and 5. Is this a right triangle?

Answer: First, the legs are lengths 5 and 8 as they are the shorter sides. 12 is the length of the hypotenuse as that is the longest side. Apply the Pythagorean Theorem and see if it works.

$a^2 + b^2 = c^2$

$5^2 + 8^2 \stackrel{?}{=} 12^2$

$25 + 64 \stackrel{?}{=} 144$

$89 \neq 144$

Conclusion: Since $a^2 + b^2 \neq c^2$, the triangle with side lengths of 5, 8, and 12 is ***not*** a right triangle.

Example Art is trying to determine whether or not the deck that he is building off of his house has square corners. (Corners that are 90 degrees are described as square corners.) He needs to make sure it is before it becomes attached to the house. He takes measurements of the length and the width of the deck. The length is 16 feet and the width is 12 feet. He measures the diagonal of the deck to be 20 feet. Does the deck have corners that form right angles?

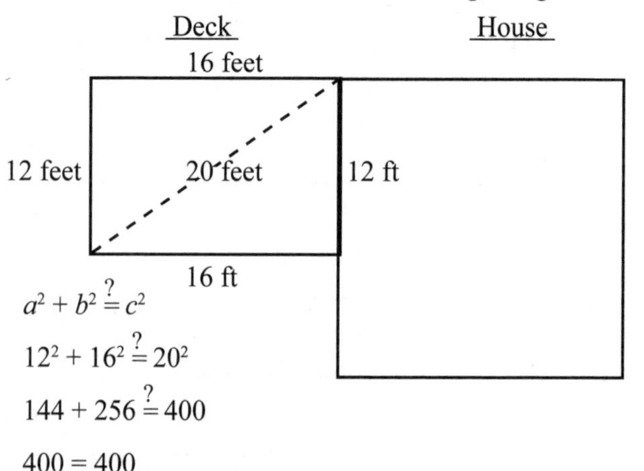

$a^2 + b^2 \stackrel{?}{=} c^2$

$12^2 + 16^2 \stackrel{?}{=} 20^2$

$144 + 256 \stackrel{?}{=} 400$

$400 = 400$

Conclusion: Since the sum of the square of the two sides equals the square of the diagonal of the deck, the deck is formed by two right triangles and has corner angles that measure 90 degrees.

NUMBERS AND QUANTITIES

- Use properties of rational and irrational numbers.
- Extend the properties of exponents to rational exponents.
- Reason quantitatively and use units to solve problems.

SUMS AND PRODUCTS OF RATIONAL AND IRRATIONAL NUMBERS

(See also Unit 1, Real Numbers and Properties of Real Numbers.)

Fractional Exponents: A fractional exponent directs us to take a specific root of a base, and then use the root as a base to be raised to a power. The denominator of the fraction indicates the root needed.

Rational Number: A real number which can be expressed as the quotient of two integers.

Irrational Number: A real number which cannot be exactly expressed as the quotient of two integers.

Some examples of irrational numbers are $\sqrt{5}$, $\sqrt[3]{24}$, $\frac{3-\sqrt{14}}{6}$, and π.

Closure: When a binary operation is performed on two members of a set and the result is always a member of that set, the set is described as having closure with regard to that operation, or being closed for that operation. The real numbers are closed with regard to addition and multiplication.

Addition of two rational numbers will result in a sum that is a rational number.

The set of integers, a subset of the real numbers, is closed with regard to addition and multiplication. If a rational number, the quotient of two integers, is added to a rational number, the sum is rational.

In this example, a, b, c, and d are all integers. The denominators, b and d are not zero. Because the set of integers has closure with regard to multiplication, ad and bc are both integers. The sum, $ad + bc$, is an integer because the set of integers is closed for addition. Therefore, the numerator in the example is an integer. The denominator is the product of two integers and is, therefore, an integer because the set of integers is closed for multiplication. Since the numerator and the denominator are both integers the definition of rational number is satisfied; the set of rational numbers has closure with respect to addition.

$$\frac{a}{b} + \frac{c}{d} = \frac{ad + bc}{bd},\ b \neq 0,\ d \neq 0$$

Examples **Adding two rational numbers.**

❶ $\frac{3}{5}+\frac{6}{7}=\frac{21+30}{35}=\frac{51}{35}$

❷ $\frac{-2}{5}+8=\frac{-2}{5}+\frac{8}{1}=\frac{-2+40}{5}=\frac{38}{5}$

❸ $\frac{0}{4}+\frac{1}{3}=\frac{0+4}{12}=\frac{4}{12}$

❹ $\frac{-5}{2}+\frac{-4}{5}=\frac{-25+(-8)}{10}=\frac{-33}{10}$

❺ $\frac{3}{4}+\frac{-6}{8}=\frac{24+(-24)}{32}=\frac{0}{32}=0$

Conclusion: The sum of two rational numbers is a rational number because their sum can be written in the form of the quotient of two integers.

Multiplication of two rational numbers will result in a product that is a rational number.

The reasoning for multiplication of two rational numbers resulting in a rational number is similar to the reasoning for addition. In the example below, a, b, c, and d are all integers. The denominators, b and d are not zero. Because the set of integers has closure with regard to multiplication, ac and bd are both integers. Since the numerator and the denominator are both integers the definition of rational number is satisfied; the set of rational numbers has closure with respect to multiplication.

$$\frac{a}{b}\times\frac{c}{d}=\frac{ac}{bd},\ b\neq 0,\ d\neq 0$$

Examples **Multiplying two rational numbers.**

❶ $\frac{3}{4}\times\frac{1}{5}=\frac{3\times 1}{4\times 5}=\frac{3}{20}$

❷ $-\frac{6}{7}\times\frac{2}{3}=-\frac{6\times 2}{7\times 3}=-\frac{12}{21}=-\frac{4}{7}$

❸ $\frac{3}{5}\times 17=\frac{3\times 17}{5\times 1}=\frac{51}{5}$

Conclusion: The product of two rational numbers is a rational number because the product can be written in the form of the quotient of two integers.

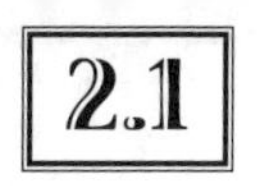

Addition of a rational and an irrational number will result in an irrational sum.

An irrational number and a rational number are in different sets. Therefore, closure does not necessarily apply. Remember that just one example that disproves a statement (called a counterexample) makes the statement false. To avoid confusion, only numerical examples are used here.

Examples **Adding a rational and an irrational number.**

❶ $2+\sqrt{5}=2+\sqrt{5}$, $\sqrt{5}$ *is irrational.*

❷ $3+\pi=3+\pi$, π *is irrational.*

❸ $\sqrt{17}+0=\sqrt{17}$, $\sqrt{17}$ *is irrational.*

Conclusion: In each case where a rational number and an irrational number are added together, the sum is irrational because the sum cannot be written in the form of the quotient of two integers.

Multiplication of a non-zero rational number and an irrational number results in an irrational product.

Examples **Multiplying a non-zero rational number**

❶ $5\left(\sqrt{2}\right)=5\sqrt{2}$, *irrational.*

❷ $-4\bullet 2\sqrt{3}=-8\sqrt{3}$, *irrational.*

❸ $100\bullet\pi=100\pi$, *irrational.*

Conclusion: In each case where a non-zero rational number and an irrational number are multiplied, the product is irrational because their product cannot be written as the quotient of two integers.

Note: **The sum or product of two irrational numbers may be irrational or it may be rational.**

Examples

❶ $\sqrt{15}+\left(-\sqrt{15}\right)=0$, *0 is rational.*

❷ $\sqrt{6}+\sqrt{10}=\sqrt{6}+\sqrt{10}$, *this is an irrational number.*

❸ $\sqrt{7}\bullet\sqrt{7}=7$, *7 is rational.*

❹ $\sqrt{5}\left(\sqrt{3}\right)=\sqrt{15}$, $\sqrt{15}$ *is irrational.*

RATIONAL EXPONENTS

(See also Unit 1 – Exponents and Radicals)

Bases and Exponents: An exponent indicates how many times to use its base as a factor. The base of the exponent is the number, variable, or parenthesis directly to the left of the exponent. If the expression to the left of the exponent is a parenthesis, then the exponent is applied to the entire content of the parenthesis. Evaluating with an exponent is often referred to as "raising" a number to a power.

Summary of Properties of Exponents:

If	=	Then
$a^x \cdot a^y$	=	a^{x+y}
$(a^x)^y$	=	a^{xy}
$\frac{a^x}{a^y}$	=	$a^{x-y}, a \neq 0$
$(ab)^x$	=	$a^x b^x$
$\left(\frac{a}{b}\right)^x$	=	$\frac{a^x}{b^x}, b \neq 0$
a^0	=	$1, a \neq 0$
a^{-1}	=	$\frac{1}{a}, a \neq 0$

Note: All the rules that apply to numbers with exponents are also used for expressions containing variables.

Examples

	Numerical	Algebraic
❶	$5^2 \cdot 5^3 = 5^5 = 3125$	$x^2 \cdot x^3 = x^5$
❷	$(2^3)^4 = 2^{12} = 4096$	$(x^2)^3 = x^6$
❸	$\frac{6^4}{6^3} = 6$	$\frac{x^5}{x^2} = x^3$
❹	$(3 \cdot 4)^2 = 3^2 \cdot 4^2 = 9 \cdot 16 = 144$	$(x^2y^3)^4 = x^8y^{12}$
❺	$\left(\frac{2}{3}\right)^2 = \frac{2^2}{3^2} = \frac{4}{9}$	$\left(\frac{x^2}{y^3}\right)^2 = \frac{x^4}{y^6}$
❻	$(5)^0 = 1$	$(x^2)^0 = 1$
❼	$5^{-1} = \frac{1}{5}$	$x^{-1} = \frac{1}{x}$

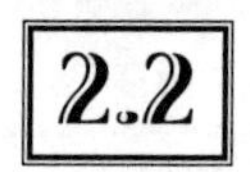

2.2

Fractional Exponents: A fractional exponent directs us to take a specific root of a base, and then use the root as a base to be raised to a power. The denominator of the fraction indicates the root needed, and the numerator tells what power to raise the root (base) to. (This process can be done in either order, but finding the root first, then raising it to a power involves numbers that are usually easier to handle.)

Reminder: The "root" of a radical sign is called the index.

Examples

❶ $8^{\frac{2}{3}} = \left(\sqrt[3]{8}\right)^2 = 2^2 = 4$

❷ $\left(\frac{25}{9}\right)^{\frac{3}{2}} = \left(\sqrt{\frac{25}{9}}\right)^3 = \left(\frac{5}{3}\right)^3 = \frac{125}{27}$

❸ $x^{\frac{4}{5}} = \left(\sqrt[5]{x}\right)^4$ *or* $\sqrt[5]{x^4}$

❹ $\left(\frac{x}{y}\right)^{\frac{1}{2}} = \sqrt{\frac{x}{y}}$ *or* $\frac{\sqrt{x}}{\sqrt{y}}$

The radical sign indicates the principal root.

Negative Fractional Exponents: The process for using negative exponents is combined here with the process for evaluating expressions with fractional exponents. Take care of the negative part of the exponent first by changing the original base to its reciprocal. Then find the root indicated by the denominator and raise it to the power of the numerator.

Examples

❶ $(64)^{-\frac{2}{3}} = \left(\frac{1}{64}\right)^{\frac{2}{3}} = \left(\sqrt[3]{\frac{1}{64}}\right)^2 = \left(\frac{1}{4}\right)^2 = \frac{1}{16}$

❷ $(-8)^{-\frac{5}{3}} = \left(\frac{1}{-8}\right)^{\frac{5}{3}} = \left(\sqrt[3]{\frac{1}{-8}}\right)^5 = \left(\frac{1}{-2}\right)^5 = \frac{1}{-32} = -\frac{1}{32}$

❸ $(x)^{-\frac{3}{2}} = \left(\frac{1}{x}\right)^{\frac{3}{2}} = \left(\sqrt{\frac{1}{x}}\right)^3$ *or* $\sqrt{\frac{1}{x^3}}$ which simplifies to $\frac{1}{x\sqrt{x}}$ *or* $\frac{\sqrt{x}}{x^2}$

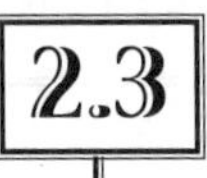

Units and Measurements

Numbers with units can be changed to different units using conversion factors. A conversion factor is a ratio (fraction) that is equal to one. We can multiply the original problem by various combinations of fractions that are equal to one until we get to the units we need.

Steps

1) Write the original problem as a fraction with the units included. If it is a whole number, put "1" with the units in the denominator.
2) Plan the conversion factors needed by determining what units the end result contains.
3) Set up the conversion factors (fractions) so that the units will cancel until you get to the units you need. Units in the numerator of one fraction should cancel with units in the denominator of another. Sometimes many factors are needed to get to the end of the problem. Cancel the units until only the units needed for the answer remain in their correct position numerator or denominator.
4) Multiply all the numerators, then all the denominators, then divide the fraction.

Quantities

Example How many inches per minute can a rabbit travel if he can go 2 miles per hour?

$$\frac{2\,\cancel{\text{miles}}}{1\,\cancel{\text{hour}}} \bullet \frac{5,280\,\cancel{\text{ft}}}{1\,\cancel{\text{mile}}} \bullet \frac{12\,\text{inches}}{1\,\cancel{\text{foot}}} \bullet \frac{1\,\cancel{\text{hour}}}{60\,\text{minutes}}$$

$$\frac{2 \bullet 5,280 \bullet 12 \bullet 1\,\text{inches}}{1 \bullet 1 \bullet 1 \bullet 60\,\text{minutes}} = \frac{2112\,\text{inches}}{1\,\text{minute}} =$$

Answer: 2112 inches per minute.

Discussion: The question is answered correctly. However, 2112 inches is a considerable number of feet. A more appropriate unit for this distance would be feet.

$$\frac{2112\,\cancel{\text{in}}}{\text{per min.}} \bullet \frac{1\text{ ft}}{12\,\cancel{\text{in}}} = \frac{176\text{ ft}}{\text{min}}$$

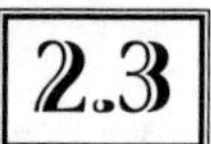

CONVERSIONS & ABBREVIATIONS

Conversions

1 inch = 2.54 centimeters
1 meter = 39.37 inches
1 mile = 5280 feet
1 mile = 1760 yards
1 mile = 1.609 kilometers

1 kilometer = 0.62 mile
1 pound = 16 ounces
1 pound = 0.454 kilograms
1 kilogram = 2.2 pounds
1 ton = 2000 pounds

1 cup = 8 fluid ounces
1 pint = 2 cups
1 quart = 2 pints
1 gallon = 4 quarts

1 gallon = 3.785 liters
1 liter = 0.264 gallon
1 liter = 1000 cubic centimeters

Metric Measurement System

10 millimeters = 1 centimeter	10 mm = 1cm
10 centimeters = 1 decimeter	10 cm = 1dm
10 decimeters = 1 meter	10 dc = 1 m
10 meters = 1 dekameter	10 m = 1 dkm
10 dekameters = 1 hectometer	10 dk = 1 hm
10 hectometers = 1 kilometer	10 h = 1 km

Commonly used conversions in the metric system include:

10 millimeters = 1 centimeter	10 mm = 1 cm
100 centimeters = 1 meter	100 cm = 1 m

In the metric system there are also conversion factors that relate length measured in meters, volume or capacity measured in liters, and weight or mass measured in grams.

1 cc (cubic centimeter) = 1 gram = 1 milliliter

1000 cc = 1 kilogram (kg) = 1 liter (1)

Rates

Conversions are often used in rate problems.

Rate is defined by this formula: $Rate = \frac{\text{Distance}}{\text{Time}}$

When using this formula it is very important to make sure the units match other parts of the problem - either other information that is given, or the units needed for the solution.

(See examples on next page)

Examples Rates

❶ What is Sam's rate of speed if he travels on his bike to a town 15 miles away from home and it takes him 2 hours?

$$R = \frac{D}{T}$$

$$R = \frac{15 \text{ miles}}{2 \text{ hours}}$$

$$R = 7.5 \text{ miles per hour}$$

❷ Using the same information as in number one, suppose we want to know the rate of speed in feet per hour instead of the miles per hour as the problem is given. It is necessary to change the units from miles to feet.

$$R = \frac{15 \text{miles}}{2 \text{hours}} \bullet \frac{5280 \text{feet}}{1 \text{ mile}}$$

Change miles to feet by multiplying using 5280 ft/1 mile. The mile units cancel leaving the units of feet which are needed for the answer.

$$R = \frac{(15)5280 \text{feet}}{2 \text{ hours}}$$

$$R = \frac{79,200 \text{feet}}{2 \text{hours}}$$

$$R = 39,600 \text{ feet per hour}$$

❸ A snail is crawling along the ground at a rate of 10 cm per minute. How many meters will he travel if he crawls for 3 hours?

$$R = \frac{10 \text{cm}}{1 \text{minute}}$$

cm needs to be changed to meters, and minutes to hours. Two conversions are needed here.

$$R = \frac{10 \cancel{\text{cm}}}{1 \cancel{\text{minute}}} \bullet \frac{1 \text{meter}}{100 \cancel{\text{cm}}} \bullet \frac{60 \cancel{\text{minutes}}}{1 \text{hour}}$$

cm and minutes both cancel leaving the rate in terms of meters/hour.

$$R = \frac{(10)(1)(60) \text{ meter}}{(1)(100)(1) \text{ hour}}$$

$$R = \frac{600\text{m}}{100\text{hr}} = \frac{6\text{m}}{1\text{hr}} = 6\text{m} / \text{hr}$$

Answer: The snail travels 6 meters in one hour so if he travels for 3 hours, he will travel 18 meters.
distance = (rate)(time)

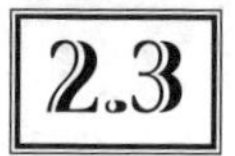

2.3

Scientific Notation: Scientific notation is used to enable people using very large or very tiny numbers in their work to write the number using a "shortcut." It uses powers of 10 which is the base of our number system as a multiplier of manageable size factors written in a standard way. The standard form of the factor is a number greater than or equal to one and less than ten. Generally two decimal places are shown. Here is the method:

Steps:

1) Move the decimal point in the number to the position immediately following the first significant digit. count the number of places the decimal is moved. (Remember that if a decimal isn't visible, it is at the end of the number.)
2) Show that the new number must be multiplied by a power of 10 with an exponent matching the number of places you had to move the decimal point.
3) Power of ten - Exponents: If the original number is greater than or equal to 10, the exponent for ten will be positive. If the original number is less than 1, the exponent is negative. (If the number was at least 1 and less than 10 to start with, the exponent is 0. It would be unusual to write a number in scientific notation with a zero exponent.)

Examples

❶ $5{,}830{,}000. = 5.83 \times 10^6$ *or* $(5.83)(10^6)$

❷ $0.000261 = 2.61 \times 10^{-4}$ *or* $(2.61)(10^{-4})$

❸

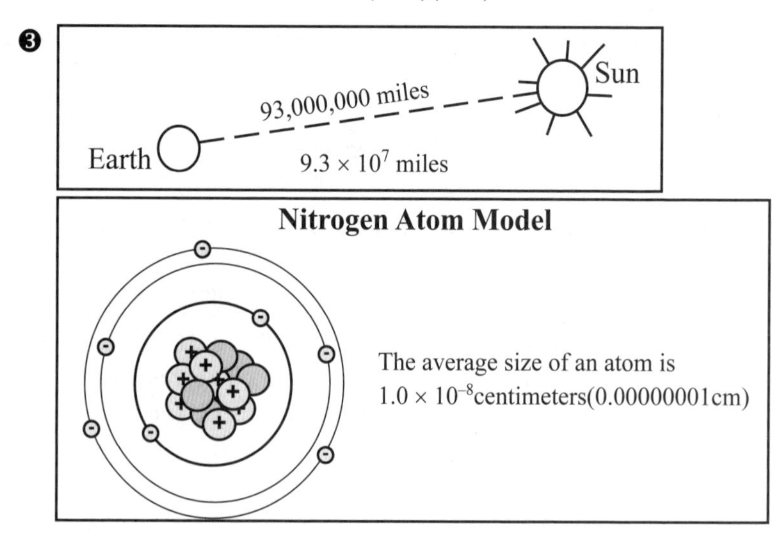

2.3

Evaluating problems using powers of 10 (scientific notation).

Steps:

1) Translate the numbers into scientific notation.
2) Use the commutative property to rearrange the numbers so the coefficients are together and the factors of 10 are together.
3) Do the appropriate operation to the coefficients and use the laws of exponents to process the powers of 10.
4) Translate back into expanded numerical form if required.

Examples

❶ $(20000)(0.00000008) = (2.0)(10^4)(8.0)(10^{-8})$

$= (2.0)(8.0)(10^4)(10^{-8})$

$= (16)(10^{4+(-8)})$

$= (16)(10^{-4})$

$= 0.0016$

❷ $\dfrac{650{,}000}{.005} = \dfrac{(6.5)(10^5)}{(5.0)(10^{-3})}$

$= (1.3)(10^{5-(-3)}) = (1.3)(10^8)$

$= 130{,}000{,}000$

❸ $\dfrac{(4000)(0.00002)}{(200{,}000)\ (0.0008)} = \dfrac{(4.0)(10^3)(2.0)(10^{-5})}{(2.0)(10^5)(8.0)(10^{-4})}$

$= \dfrac{(4.0)(2.0)(10^3)(10^{-5})}{(2.0)(8.0)(10^5)(10^{-4})}$

$= \dfrac{(8.0)(10^{-2})}{(16.0)(10)}$

$= (0.5)(10^{-2-1}) = (0.5)(10^{-3})$

which is equivalent to $(5.0)(10^{-4})$

$= 0.0005$

Quantities

ERROR IN MEASUREMENT

Note: Relative error is included in this section for reference, in order to have a more complete understanding of measurement.

Actual (or True) Measure: The measure of a quantity that is accepted as being accurate and correct. This might be obtained from a chart, reference table, or other acceptable resource.

Experimental Measure: The measure of a quantity obtained by the person performing the measurement.

Example A student measures the mass of an object to be 156 grams. The actual mass is 173 grams.
– Actual Measure is 173 grams.
– Experimental Measure is 156 grams.

Absolute Error: The actual physical difference between the actual measure and the experimental measure of the item. The absolute error is the absolute value of the difference between the actual measure and the experimental measurement. It is a positive number and is labeled with whatever units are used in the measure.

– Actual Measure – Experimental Measure = $E_{absolute}$
– Absolute Error: 173 g – 156 g = 17 g

Relative Error: The relative error shows the importance of the error in relation to the size of the object measured. When we have an accepted value given in the problem, use this formula to find the relative error:

$$E_{relative} = \frac{E_{absolute}}{M_{actual}} \qquad E_{relative} = \frac{17\text{ g}}{173\text{ g}} = 0.098265896$$

We rounded this number to the nearest 1000th for convenience.

$$E_{relative} = 0.098$$

Percent of Error: Often the relative error is expressed in % form.

$$0.098 = 9.8\%$$

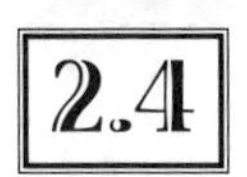

Ratio of Absolute Error to Actual Measure in Different Circumstances:

The quantity of the absolute error has different meaning in problems with various actual measurements. Use relative error to express accuracy as a relative quantity.

In another experiment, the absolute error might again be 17 grams just as it was in our primary example. However, if the actual measure is smaller than the example above, the relative error and percent of error would be considerably higher.

Example If the actual measure is 75 g, the relative error is

$\frac{17g}{75g} \approx 0.227$ and the percent of error would be 22.7%.

On the other hand, if the actual measure is larger than the 173 g in our original experiment, the relative error is smaller and the percent of error is less.

Example If the actual measure is 300 g, the relative error is

$\frac{17g}{300g} \approx 0.057$ which is only 5.7% of error.

Quantities

Examining Error in Problem Solving

When quantities that involve a measurement error are used to calculate another answer, an even larger error can result. So when a measurement with a small relative error is used in a subsequent calculation, the final answer often involves a much larger relative error.

Example Using a ruler, Graeme measured the sides of a rectangular prism. His measurements were L = 5.2 cm, W = 8.3 cm, and the height was 4.2 cm. The actual correct measurements of the rectangular prism were L = 5.0 cm, W = 8.0 cm, and H = 4.0.

Linear

Percent of error in Length = $\frac{5.2\,\text{cm} - 5.0\,\text{cm}}{5.0\,\text{cm}} = 0.04 = 4\%$

Percent error in Width = $\frac{8.3\,\text{cm} - 8.0\,\text{cm}}{8.0\,\text{cm}} = 0.0375 = 3.75\%$

Percent error in Height = $\frac{4.2\,\text{cm} - 4.0\,\text{cm}}{4.0\,\text{cm}} = 0.05 = 5\%$

Note: Compare both the actual and relative errors in length to the same error in height.

Area

Area of one side using Graeme's measurement:

$$A = LW$$
$$A = (5.2)(8.3) = 43.16\ \text{cm}^2$$

Area using correct measurements:

$$A = LW$$
$$A = (5)(8) = 40\ \text{cm}^2$$

Absolute error: $43.16 - 40\ \text{cm} = 3.16\ \text{cm}$

Relative Errorr: $\dfrac{3.16\ \text{cm}}{40\ \text{cm}} = .079$

Percent of error: 7.9%

Notice that percent of error for the area is larger than the percent of error in the linear measurement.

Volume

Volume of the rectangular solid using Graeme's measurements:

$$V = LWH$$
$$V = (5.2)(8.3)(4.2) = 181.272\ \text{cm}^3$$

Volume using correct measurements

$$V = LWH$$
$$V = (5.0)(8.0)(4.0) = 160.0\ \text{cm}^3$$

Percent of Error in Volume $\dfrac{181.272 - 160}{160} = 0.13295 = 13.295\%$

Conclusion: Although the original percent of error in measurement may not seem high, when the measurements are used in the area formulas, the percent of error increases. When they are used in the volume, the percent of error is significantly higher, it is more than tripled in this particular example!

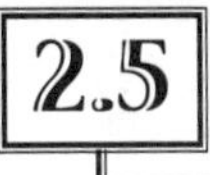
2.5

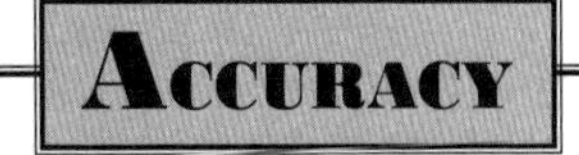
Accuracy

Precision and Accuracy: When measuring quantities, there are two words that are commonly used. Levels of certainty about the correctness of the measurement depend on the ability of the measuring tool to be appropriate to the task at hand.

Note: When working with data, the greatest precision for a result is only at the level of the least precise data point. (e.g. If units are tenths and hundredths, then the appropriate level of precision is tenths.)

Quantities

Precision: The measurement of a quantity is repeated as long as there is no change in conditions. For example, you are weighing an object in the lab and each time you weigh it the scales register 5.5 kg. This shows a precise measurement.

Accuracy: Accuracy is determined by how close a measured quantity is to the actual, known value. For example, you weigh an object known to have a weight of 5.5 kg and you get a reading of 4.2 kg for that substance, then your measurement is not accurate. The result of your measurement is not close to the actual value. Analysis of each situation is needed to determine what degree of accuracy is acceptable.

Examples

❶ At Dan's delicatessen, the scale is regulated and tested for precision regularly to insure that it is measuring the meat or cheese correctly to the nearest hundredth of a pound. Dan puts an unopened roll of salami labeled as 5 lbs, on the scale. The scale indicates the roll is 4.75 pounds, a quarter of a pound less than it should be. After double checking his measurement, Dan complains to the supplier that he is not getting the quantity of salami for which he paid. Since the scale in the delicatessen has been recently tested and certified as being precise, the meat supplier's labeling of the 5 pound roll of salami was not accurate as it did not match the actual weight of the salami.

Dan's customers often ordered ¼ pound of salami. The level of precision of the scales, the hundredth of a pound, is important. If whole pounds had been used in this example, the measurement would have been interpreted as 5 pounds and the delicatessen owner would not have realized he was ¼ pound short on his order. Perhaps the meat supplier used a scale that was precisely measured to the nearest pound – Dan's order could have contained anywhere between 4.5 and 5.5 pounds of meat! Or perhaps the supplier's scale was not tested for precision as Dan's was, and therefore an inaccurate measurement was obtained.

❷ In a science experiment, the mass of material is measured by Tim and Joe using the same sample. If the mass is 654.8 mg on Tim's measurement and 701.2 mg on Joe's, the accuracy of the measurements is in question. It is necessary to determine the actual mass of the sample in order to discuss the accuracy of the measurements. If both students did their work on the same scale, the accuracy may have been impacted by the placement of the sample on the scale, or perhaps the interpretation of the reading* by Joe was different than by Tim. If different scales were used, then one scale is more accurate than the other. If the actual measure of the mass is 700.5 mg, the measurement made by Joe is more accurate. Since the actual mass is known to be 700.5 mg to the nearest tenth of a milligram, the boys using the same level of accuracy, tenth of a milligram, is appropriate.

* With digital scales, the readings are quite clear, but in scales with mechanical balances, the reading may be open to interpretation.

❸ Margo and Peter are working on averaging the volume of various samples of a liquid measured by other students in their science class. What is the average volume of the 5 samples?

The following are the measurements they were given:

Sample	Volume in cc's
A	10.23
B	8.725
C	11.4
D	10.057
E	9.25

Solution: Because the least precise data point is measured in tenths, the appropriate level of precision in the answer is to the tenths place.

$$\frac{10.23 + 8.725 + 11.4 + 10.057 + 9.25}{5} = 9.9324 = 9.9\ cc$$

Level of Accuracy: In both examples above, the levels of accuracy can be compared by using the Error in Measurement described on page 36. Additionally, the level of accuracy depends on the measuring tool being used and on the size of the item being measured.

Appropriate Units of Measure: Both examples above involved measures that used unit of pounds or milligrams. The scales used were appropriate. To use a scale that measured in tens of pounds (example 1) or kilograms (example 2) would not allow for measurements close to the actual measurements.

In other types of problems, it may be reasonable to use tons for the unit of measure, or perhaps micrograms. A cruise ship would be measured by the ton. Components in a medicine might be measured by micrograms.

In measuring distance, if the need is to be accurate about how many miles it is between two cities, the accuracy may be acceptable to the nearest mile if they are close. If they are very far from each other, distance might be measured in tens of miles, or hundreds of miles. There isn't usually a need to be accurate to the nearest mile if we are discussing the distance from Los Angeles to New York, hundreds of miles (or maybe even thousands) would be accurate enough.

Precision, accuracy, appropriate units of measure, and error in measurement are all related topics. Each experiment, description, or discussion requires that the situation be analyzed to make decisions that make sense for the situation.

EXPRESSIONS AND EQUATIONS

- Interpret the structure of expressions.
- Create equations that describe numbers or relationships.
- Understand solving equations as a process of reasoning and explain the reasoning.
- Solve equations and inequalities in one variable.
- Write expressions in equivalent forms to solve problems.
- Solve systems of equations.
- Perform arithmetic operations on polynomials.

3.1 Algebraic Language

Term: It indicates a product and may have one factor or many factors. It may contain a numerical factor called a coefficient, and one or more variables which may have exponents. A coefficient or an exponent of "1" is not written.

Example x, $3x^2$ *or* xy

Coefficient: A numerical factor written to the left of a variable. A coefficient of one is understood and is not written.

Example $6x^2$: the coefficient is 6; *or* y^3 : the coefficient is 1.

Constant: A coefficient without a base and exponent. It's just a number.

Example $y = 5x + 7$, 7 is a constant.

Factor: A number or expression that is multiplied by another to result in a particular product.

Example $(x - 5)(x + 2)$ are factors of $x^2 - 3x - 10$.
$x(x + 10)$ are factors of $x^2 + 10x$.

Like Terms: Have the same LETTER BASE and the SAME EXPONENT. They can be combined by adding the numerical coefficients (with their signs) and keeping the base (variable) and its exponent UNCHANGED.

Example $5x^3 + 4x^3 = 9x^3$ *or* $x^2 - 6x^2 = -5x^2$

Unlike Terms: Have different variables and/or different exponents. These **cannot** be *combined* or *added.* (Unlike terms can be multiplied when indicated.)

Example The following cannot be simplified; $3x + 4$, $x^2 + 4x$, *or* $x^2 + y^2$

Expression: Two or more terms, with or without a constant, that are connected with + or –.

Example $x + y$, $2x^2 + 5$, $6xy + 7$, $3x^2 + 2x + 4$

Equation: An expression that is equal to a constant or to another expression and is connected using an = sign.

Example $x = 5$, $x + 3 = 2y$, $x^2 - 6x - 7 = 0$

Function: An equation with specific characteristics. See Unit 4.

Polynomial: Terms are separated by + or – to form polynomials. Monomials, binomials, and trinomials are polynomials with special characteristics.

1) **Monomial:** A *single algebraic term* or the *product* of algebraic terms.

   Example $4x^2$ *or* $12xy$ *or* $8x^2yz^2$ *but not* $\frac{3x}{2y}$

2) **Binomial:** The sum or difference of 2 (unlike) monomials that cannot be combined.

   Example $4x^2 + 12x^2y$ *or* $5m - 2$ *or* $x^2 - 4$

3) **Trinomial:** The sum or difference of 3 monomials that cannot be simplified or combined.

   Example $3x + 4y + 8$ *or* $x^2 + 4x + 5$

STANDARD FORM

Quadratic functions are in the form $f(x) = ax^2 + bx + c$, $a \neq 0$ which is called standard form. Standard form is needed in order to facilitate factoring, graphing, and the use of the quadratic formula.

Functions of higher degrees, with larger exponents, are considered to be in **standard form** when the variable with the highest exponent is listed first, followed by that variable with decreasing exponents until the constant is reached. The highest exponent is referred to as n, and each decreasing exponent is $n-1, n-2$, etc. When we get to $n-n$, that equals an exponent of zero, $n^0 = 1$.

Examples ❶ $f(x) = x^4 - x^3 + x^2 - x + 1$

❷ $g(x) = 3x^5 + 2x^4 - 5x^3 - x^2 + 5$

Structure of Expressions

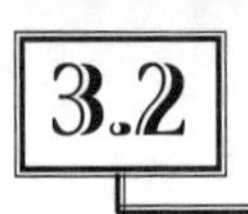

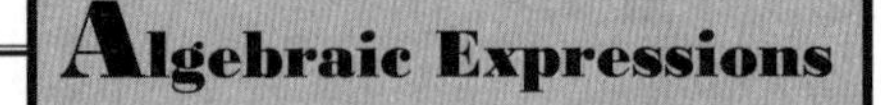

3.2 Algebraic Expressions

Identifying Parts of Algebraic Expressions: It is necessary to identify individual parts of complicated expressions in order to analyze the method needed for proceeding with a problem.

Examples

1. $5x^2$: 5 is the coefficient, x is the variable. Additionally 5 and x are both factors of the term. 2 is the exponent that is to be applied to x. Other ways to write this might be $5(x^2)$ or $5(x)(x)$.
2. $6x^2 + 7$: This is a binomial expression. $6x^2$ and 7 are both terms of the expression. 7 is a constant. 2 is an exponent applied to the variable, x. 6 is the coefficient of the variable.
3. $4(3x^2 + 5)$: 4 is a factor in this expression. $(3x^2 + 5)$ is a binomial factor.
4. $5x\,(x + 2)^2$: $5x$ is a factor. $(x + 2)$ is a binomial factor that is to be raised to the 2nd power. 2 is the exponent and it is applied only to the binomial factor $(x + 2)$.
5. $(x + 1)(x - 2)^2$: $(x + 1)$ is a binomial factor. $(x - 2)$ is another binomial factor. The exponent of 2 is to be applied only to $(x - 2)$.
6. $R(x - 1)^t$: R and $(x - 1)$ are both factors and t is the exponent that is applied to $(x - 1)$. Multiply $(x - 1)$ by $(x - 1)$ t times. Then multiply that answer by R.

Equivalent Forms of Expressions: It is often necessary to rewrite an expression or equation in order to solve problems. The properties and laws of real numbers also apply to polynomials. Sometimes expanding the expression is appropriate and in other cases simplifying the expression is needed. The following are some examples written in equivalent forms.

Examples

1. $(x + 2)(x - 5)$ can be written as $x^2 - 3x - 10$.
2. $\sqrt{9x^3}$ is equivalent to $3x\sqrt{x}$.
3. $5x(x + 2)^2$ can be multiplied out and written $5x(x^2 + 4x + 4)$ which is equivalent to $5x^3 + 20x^2 + 20x$.
4. $6x^2 + 12x - 18$ is equivalent to $6(x^2 + 2x - 3)$ which can also be written $6(x + 3)(x - 1)$.
5. $x^4 - y^4$ is the difference of 2 perfect squares. It can be rewritten as $(x^2 - y^2)(x^2 + y^2)$. Since $(x^2 - y^2)$ is still the difference of two perfect squares, this can be factored further as $(x - y)(x + y)(x^2 + y^2)$.

(See also Factoring page 49, 95)

3.3

Equivalent Forms of Formulas

Formulas are literal equations. They often need to be written in equivalent forms. The directions will usually say, "Solve for t in terms of r and s." or sometimes just, "Solve for m." Solving means to isolate the variable designated. The answer will still have the other letters (and sometimes numbers) in it.

Example $A = lw$ Directions will usually say: Solve for "w" in terms of A and l.

Steps **1)** Divide both sides by l to isolate w: $\frac{A}{l} = \frac{lw}{l}$

2) Answer is: $\frac{A}{l} = w$ *or* $w = \frac{A}{l}$

Think of the letter you are solving for as the variable in any ordinary equation. Use the same method you would use to solve a regular equation: Isolate the variable. The answer will contain letters, showing the variable isolated and separated from its coefficient. All the other letters will be on the other side of the equal sign.

Example Solve for b_1 in terms of the other variables when given the formula for the area of a trapezoid.

$A = \left(\frac{b_1 + b_2}{2}\right)h$ Formula as given.

$A = \frac{hb_1 + hb_2}{2}$ Distribute the h.

$2A = hb_1 + hb_2$ Multiply both sides by 2 to remove the fraction.

$-hb_2 \qquad -hb_2$ Subtract hb_2 from both sides to isolate the term containing b_1.

$\frac{2A - hb_2}{h} = \frac{hb_1}{h}$ Divide both sides by h to isolate b_1.

$\frac{2A - hb_2}{h} = b_1$ *or* $b_1 = \frac{2A - hb_2}{h}$ Solution.

Note: There are other algebraic methods that can be used to isolate the desired variable. Be able to explain your method if you are asked to do so.

Creating Equations

3.3

FORMULAS FOR CIRCLES

Use: $\boldsymbol{d = 2r}$ to find the diameter if you have the radius.

Use: $\boldsymbol{r = d/2}$ to find the radius if you know the diameter.

Use: $\mathbf{C = 2\pi\, r}$ or $\mathbf{C = \pi d}$ to find the circumference.

Use: $\mathbf{A = \pi\, r^2}$ to find the area of a circle.

To find the radius or diameter if you have the circumference, substitute the circumference in the formula and solve for r. Use the same process if you have the area - substitute the value you have for area in the area formula and find the radius. Remember that the formula has r^2 in it, so make sure you find the square root of r^2. Leave π as a symbol.

π is the ratio of the circumference of a circle to its diameter. It is an irrational number. Leave the answer in terms of π unless directed otherwise.

Follow the directions!
You will lose credit on a problem if it says "leave in terms of π" or "find the exact answer" and you do not leave π as a symbol. π is a more exact answer than calculations performed using its approximate value of 3.14 or $\frac{22}{7}$.

3.4 Factoring

Factors: Numbers, terms, or expressions that are multiplied together to form a product. A polynomial inside a ( ) is a factor if the ( ) indicates multiplication. not just grouping.

Examples

❶ $4 \cdot 3$; 3 and 4 are both factors of 12.

❷ $5x\,(2x)$ means $5 \cdot x \cdot 2 \cdot x$; 5, x, and 2 are all factors of $10x^2$.

❸ $(x + 2)(x - 3)$; $(x + 2)$ and $(x - 3)$ are binomial factors.

❹ $(x - 2) + (x + 3)$; These ( ) are used for grouping to show addition of two binomials. The binomials are not factors.

Factoring An Expression: To factor an expression, break it down into its prime factors. There will only be one correct set of prime factors for any expression.

Prime Factors: These are the factors of a product that is broken down as far as possible while still resulting in the same product. The prime factors of a number or of an algebra problem will always be the same.

Examples

❶ $12 = 2 \cdot 6$; $12 = 2 \cdot 2 \cdot 3$; 2, 2, and 3 are all prime factors of 12.

❷ $3x + 6 = 3(x + 2)$; 3 and $(x + 2)$ are the prime factors of $3x + 6$.

❸ $12xy + 3x$: $3x\,(4y + 1)$; 3 and x and $(4y + 1)$ are prime here.

Common Factor (GCF): A factor which is present in each term in an expression. The greatest common factor or GCF is the largest factor that is present in each of the terms to be considered. Each term in the expression must be divisible by the same number(s) and/or variable(s) if a GCF exists. A GCF greater than one does not always exist. If it does, we "factor it out" which means to divide each term in the expression by the GCF. It is kept with the other factors.

Examples

❶ GCF of $4x + 8y$ is 4; $4x$ and $8y$ are both divisible by 4. Factors are $4(x + 2y)$.

❷ GCF of $x^2 + 2x$ is x; x^2 and $2x$ are both divisible by x. Factors are $x(x + 2)$.

❸ GCF of $12xy - 4xyz$ is $4xy$; $12xy$ and $-4xyz$ are both divisible by $4xy$. Factors are $4xy(3 - z)$.

General Procedure for Factoring:

Steps

1) In any factoring problem, put the terms in standard form first.
Note: In an equation, make one side of the equation = 0 by moving all the variables and numbers to one side of the equal sign. (Use the usual algebraic methods of adding or subtracting terms from both sides of the equation.)

2) LOOK at each term in the problem to see if there is a GCF. If there is, factor out the GCF and show it at the left side of the remaining expression. The GCF remains as part of the problem. Go to step 3. If there is no GCF, go directly to step 3.

3) Now look INSIDE the parenthesis, or just at the problem itself if there was no GCF, to see if what is left can be factored. Is it a binomial or trinomial?
(See next page to factor a binomial, and see page 52 if it is a trinomial.)

Examples

❶ $2x^2 + 6x - 8$ GCF is 2.
$2(x^2 + 3x - 4)$ Show the 2 at the left, and the quotient after "factoring out" the 2 in parenthesis.

❷ $x^2 + 2x - 15$ No GCF. Go to step 3

❸ $2x^3 - 8x$ GCF is $2x$.
$2x(x^2 - 4)$ $2x$ and the quotient are both shown.

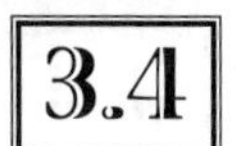

FACTORING BINOMIALS

The Difference of Two Perfect Squares – like $x^2 - a^2$. This can be factored into $(x - a)(x + a)$. When $(x - a)(x + a)$ are multiplied together the result is $x^2 - a^2$.

Examples

❶ $x^2 - 9$
$(x + 3)(x - 3)$
This expression has no GCF. It is the difference of two perfect squares.

❷ $2x^3 - 8x$
$2x(x^2 - 4)$
$2x(x - 2)(x + 2)$
This expression has a GCF, and the quotient is the difference of two perfect squares.

❸ $4x^2 - 81$
$(2x + 9)(2x - 9)$
In this example, the first term has a perfect square number as well as a squared variable and the last term in a perfect square. It is the difference of two perfect squares.

The Sum of Two Perfect Squares – like $x^2 + a^2$. This is prime and cannot be factored.

Examples

❶ $x^2 + 16$
Both terms are perfect squares and they are added. This cannot be factored.

❷ $2x^2 + 18$
$2(x^2 + 9)$
There is a GCF of 2. Factor that out and the quotient that remains is prime.

Some Binomials are Not Factorable – some have a GCF and that is all that can be factored, others are prime as they are.

Examples

❶ $3x^2 + 6$
$3(x^2 + 2)$
GCF of 3.
Cannot be factored further.

❷ $5x^2 - 4$
Examples 2, 3, and 4 cannot be factored at all.

❸ $x^2 + 1$

❹ $x^2 - 2$

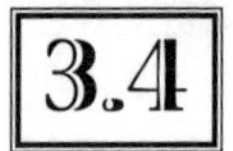

Factoring Trinomials into 2 Binomial Factors: When working with a trinomial in standard form, we use "first", "middle", and "last" when referring to the position of the term. For example, in the trinomial: $5x^2 + 3x - 2$, $5x^2$ would be the first term, $3x$ is the middle term, -2 is the last term and is also called the constant. As a reminder, the standard form of a quadratic trinomial is $ax^2 + bx + c$. The numbers represented by a, b, and c, can each be negative or positive real numbers.

The *leading coefficient, "a"*, is the numerical factor of the first term. In the example above, $5x^2$ is the first term which makes 5 the leading coefficient. When written in standard form, a quadratic trinomial's leading coefficient is usually referred to as "a".

> ***Note***: If the leading coefficient is one, it is not written in front of the first term. In algebra, the number one is rarely written down when it is used to multiply something. x^2 means $(1)(x^2)$.

The coefficient of the 2nd term, the x term, is called "b". Again, if it is one, it is not written down. x means $(1)(x)$.

The constant, or the third term which is a number only, is called "c." If c is 1, then "1" **IS** written here because it is not a factor but is simply a number.

Follow these steps to successfully factor completely a quadratic trinomial into its binomial factors. (***Note***: Not all quadratic trinomials can be factored and we have other methods of working with them that will be presented at a later time.) (See page 97)

Steps

1) Make sure the trinomial is in standard form, $ax^2 + bx + c$, and that a GCF has been factored out if there is one. Remember that the GCF stays in the problem.
2) Identify the values of b and c, including their + or – signs. On scrap paper, write down all the pairs of factors (numbers) that multiply together to give a product equal to "c".
3) Find the sum (add them together) of each pair of factors for "c" that you found in step 2 and identify the pair that has a sum equal to "b" - making sure the sum has the correct sign. Only one pair will work correctly.
4) Underneath the original trinomial, make 2 sets of parentheses, side by side. Each parenthesis will contain a binomial factor when we are finished.
5) In both parentheses, put x as the first term. The 2nd terms in the each parenthesis will be the pair of factors (from step 4) of "c" whose sum equals "b" (from step 5). Include the positive or negative signs.
6) You can check to make sure the factors are correct by multiplying the 2 binomials back together. If a GCF was found, make sure to multiply by it back in using the distributive property as the last step. The original trinomial should be the result.

3.4

Binomial Factors of a Trinomial: Standard form: $ax^2 + bx + c$

Steps:

1) Make two sets of parenthesis. The factors will be binomials.
2) In the first term of each ( ), a number and the variable will be written. For the 2nd term in each, a number will be written.
3) The product of the coefficients of the first two terms must be a. The product of the last two terms must be c. When these two products are added, they must equal b.
4) When the two binomials factors are multiplied back together, the product is $ax^2 + bx + c$.

Note: in some problems the 2nd term in each factor may contain a variable.

For example: $x^2 + 3xy + 2y^2$

$(x + 2y)(x + y)$

- When the leading coefficient, $a = 1$: The variable in the factors will have 1 as a coefficient in both parenthesis. The 2nd terms in each ( ) have a product of c and a sum of b as long as both terms are only numbers.

Examples

❶ $x^2 - 3x - 4$
$(x - 4)(x + 1)$

❷ $x^2 + 8x + 15$
$(x + 5)(x + 3)$

❸ $x^3 - 3x^2 + 2x$
$x(x^2 - 3x + 2)$
$x(x - 2)(x - 1)$

- When a does not = 1, try ARC factoring. Although factoring can be done by trial and error or by several other methods, this method works well for my students.

Note: Follow your teacher's directions about factoring.

Examples

❶ When the leading coefficient is not 1: $2x^2 + 15x + 18$

Steps:

1) Factor out a GCF if possible. (none here)
2) Multiply $(a)(c)$: $2x^2 + 15x + 18$
3) Rewrite with $a = 1$ and $c = ac$. $x^2 + 15x + 36$ *ARC*
4) Factor the remaining trinomial. $(x + 12)(x + 3)$
5) In both ( ), divide the 2nd terms by the original a. $\left(x + \frac{12}{2}\right)\left(x + \frac{3}{2}\right)$
6) Simplify if possible. Reduce fractions. $(x + 6)\left(x + \frac{3}{2}\right)$
7) If a fraction remains in one or both parentheses, its denominator becomes the coefficient of x in that parenthesis. $(x + 6)(2x + 3)$

9) These are the factors of the original trinomial. $(x + 6)(2x + 3)$

Remember: If a GCF was found, that remains a factor in the final answer.

Seeing Structure in Expressions

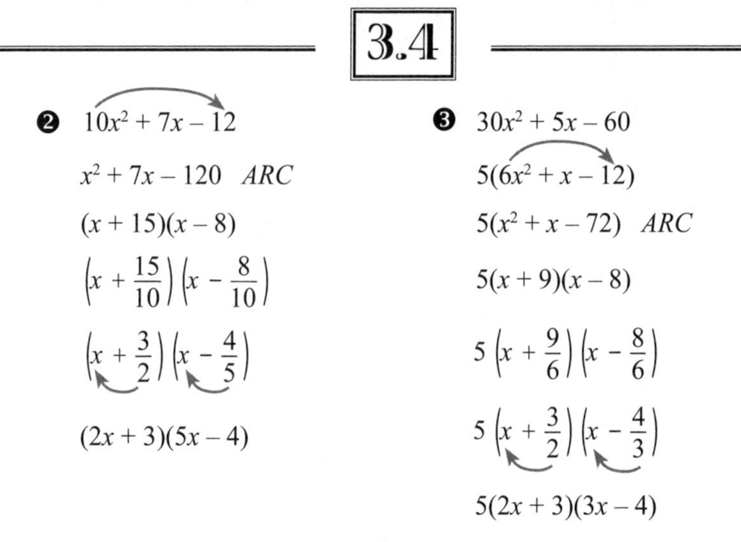

ARC refers to the connections between the terms as shown in the examples.

Factoring by Completing The Square: This method of factoring can be used to solve some quadratic equations that cannot be readily factored. The examples used here to demonstrate the factoring technique are expressions, not equations. The first step in learning to use completing the square as a solution method is to determine how to make an expression into a perfect square trinomial which can be factored.

- **Create the perfect square trinomial:** In an expression $\boldsymbol{ax^2 + bx}$, a value of $\boldsymbol{c}$ can to be added to make it into a perfect square trinomial which can then be factored. The required value of $\boldsymbol{c}$ can be found by using the values of $\boldsymbol{a}$ and $\boldsymbol{b}$. In order to us this method, the coefficient of the squared term of the expression must be 1. Sometimes it is necessary to divide by $\boldsymbol{a}$ to achieve that result. (See examples 2 and 3 on pages 55 and 56.)

- **Factor:** The final values of $\boldsymbol{a}$ and $\boldsymbol{c}$ in the perfect square trinomial are used to find the factors. When completed and written in the form $\boldsymbol{ax^2 + bx + c}$ the factors of the perfect square trinomial are $\left(\sqrt{ax^2} + \sqrt{c}\right)^2$. When simplified, the factors are $\left(\sqrt{a}x + \sqrt{c}\right)^2$.

Examples In these expressions find the number needed to complete the square, write the perfect square trinomial, and write the factors of the perfect square.

(Examples are on the next two pages.)

❶ $x^2 + 20x$

Steps:

1) $x^2 + 20x$ is in standard form: $a = 1, b = 20.$

2) Divide b by 2 and square it: $\left(\frac{b}{2}\right)^2 = \left(\frac{20}{2}\right)^2 = 10^2 = 100$

3) Add the value of $\left(\frac{b}{2}\right)^2$: to the expression $x^2 + 20x + 100$

4) Factor using $\left(\sqrt{ax^2} + \sqrt{c}\right)^2$: $\left(\sqrt{1x^2} + \sqrt{100}\right)^2$

5) Simplify the factors: $(x + 10)^2$

Answers:

- The number needed to complete the square is 100.
- The perfect square trinomial is $x^2 + 20x + 100$.
- The factors are $(x + 10)(x + 10)$ *or* $(x + 10)^2$

❷ $3x^2 + 6x$

Steps:

1) $3x^2 + 6x$ is in standard form: $a = 3, b = 6$. (***Note:*** $a \neq 1$)

2) Since a is a factor of b, a is the greatest common factor (GCF): $3(x^2 + 2x)$

3) Now $a = 1$ and $b = 2$. Divide b by 2 and square it: $\left(\frac{b}{2}\right)^2 = \left(\frac{2}{2}\right)^2 = 1$

4) Add the value of $\left(\frac{b}{2}\right)^2$ to the expression inside the parenthesis: $3(x^2 + 2x + 1)$

5) Distribute the common factor: $3x^2 + 6x + 3$

6) Factor using $\left(\sqrt{ax^2} + \sqrt{c}\right)^2$: $\left(\sqrt{3x^2} + \sqrt{3}\right)^2$

7) Simplify: $\left(\sqrt{3}x + \sqrt{3}\right)^2$

Answers:

- The number needed to complete the square is 3.
- The perfect square trinomial is $3x^2 + 6x + 3$
- The factors are $\left(\sqrt{3}x + \sqrt{3}\right)\left(\sqrt{3}x + \sqrt{3}\right)$ *or* $\left(\sqrt{3}x + \sqrt{3}\right)^2$.

❸ $8x^2 + 5x$

Steps:

1) $8x^2 + 5x$ is in standard form: $a = 8,\ b = 5$ (***Note:*** $a \neq 1$)

2) Although a is **not** a factor of b, divide the expression by a: $8\left(x^2 + \frac{5}{8}x\right)$

3) Now $a = 1$ and $b = \frac{5}{8}$. Divide b by 2 and square it: $\left(\frac{b}{2}\right)^2 = \left(\frac{5}{8} \div 2\right)^2 = \left(\frac{5}{16}\right)^2 = \left(\frac{25}{256}\right)$

4) Add the value of $\left(\frac{b}{2}\right)^2$ to the expression inside the parenthesis: $8\left(x^2 + \frac{5}{8}x + \frac{25}{256}\right)$

5) Distribute: $8x^2 + 5x + \frac{25}{32}$

6) Factor using $\left(\sqrt{ax^2} + \sqrt{c}\right)^2$: $\left(\sqrt{8x^2} + \sqrt{\frac{25}{32}}\right)^2$

7) Simplify:

$$\left(2\sqrt{2}x + \frac{\sqrt{25}}{\sqrt{32}}\right)^2 = \left(2\sqrt{2}x + \frac{5}{4\sqrt{2}}\right)^2 = \left(2\sqrt{2}x + \frac{5 \bullet \sqrt{2}}{4\sqrt{2} \bullet \sqrt{2}}\right)^2 = \left(2\sqrt{2}x + \frac{5\sqrt{2}}{8}\right)^2$$

Answer:

- The number needed to complete the square is $\frac{25}{32}$.
- The perfect square trinomial is $8x^2 + 5x + \frac{25}{32}$.
- The factors are $\left(2\sqrt{2}x + \frac{5}{4\sqrt{2}}\right)^2$ or $\left(2\sqrt{2}x + \frac{5\sqrt{2}}{8}\right)^2$.

Factors and Zeros:

Recall that on the x axis, $y = 0$. When an equation is equal to zero and the equation is factored, or solved with the quadratic formula, the zeros of the equation, also called the roots, are found. The value(s) of x are the points of intersection between the graph of the equation and the x-axis. The zero product property tells us that if any factor in a product is zero, then the product is zero. To find the zeros of the equation, make each factor equal to zero and solve for x. This enables us to locate on a graph the x-intercepts. Although more information is needed, a rough sketch of the graph can be made.

Example Find the zeros and sketch a graph of the equation.

$x^2 - 4x - 5 = y$

Make $y = 0$, $x^2 - 4x - 5 = 0$

Factor $(x - 5)(x + 1) = 0$

Solve $x - 5 = 0 \quad x + 1 = 0$

$x = 5 \quad x = -1$

More information is needed before sketching the graph, but we know that it intersects the x-axis at 5 and at -1. To make a more educated sketch, remember that $x = 0$ on the y-axis. Substitute 0 for x and solve the original equation for y. (See pages 123-126 for additional information about graphing.)

$y = x^2 - 4x - 5$

$y = 0^2 - 4(0) - 5$

$y = -5$

The graph intersects the y-axis at -5. The graph of a quadratic equation is a parabola. This parabola crosses the y-axis at -5, and the x-axis at 5 and -1. Here is a graph of $y = x^2 - 4x - 5$:

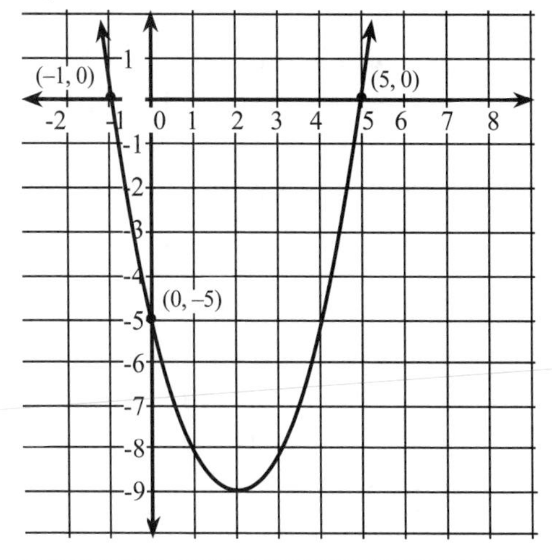

Arithmetic with Polynomials & Rationals

Operations With Polynomials

The properties and laws of real numbers can be applied to polynomials.

Closure: Polynomials are closed under the operations of addition, subtraction, and multiplication.

Addition: Remove any parenthesis using the distributive property if needed. (You can always put a "1" in front of a ( ) if it is easier for you.) Then combine like terms.

Example $2(x^2 + 3x - 4) + 3x^2 - 4x$

Steps **1)** Distribute the 2: $2x^2 + 6x - 8 + 3x^2 - 4x$

2) Collect like terms: $5x^2 + 2x - 8$

Example $x^3 + 2x - 3 + 3(x^2 - 5)$

Steps **1)** Distribute the 3: $x^3 + 2x - 3 + 3x^2 - 15$

2) Collect like terms: $x^3 + 3x^2 + 2x - 18$

Subtraction: Multiply the terms in the polynomial to be subtracted by –1 (use the distributive property). This removes the parenthesis and takes care of the SIGN CHANGES. Then ADD by combining like terms.

Example $(3x - 2) + (5x - y) - (2x - 4)$

Steps **1)** Put in "1's": $1(3x - 2) + 1(5x - y) - 1(2x - 4)$

2) Use the distributive property: $3x - 2 + 5x - y - 2x + 4$

3) Simplify or collect like terms: $6x - y + 2$

Example From the sum of $(x + 2y)$ and $(2x - y)$, subtract $(3x - y)$.

Steps **1)** Put brackets around the sum first: $[(x + 2y) + (2x - y)] - (3x - y)$

2) Simplify inside the bracket: $[x + 2x + 2y - y] - (3x - y)$

3) Put in "1's": $1(3x + y) - 1(3x - y)$

4) Use the distributive property: $3x + y - 3x + y$

5) Simplify or collect like terms: $2y$

Example Subtract $5x - 2y$ from $12x - 5$

Steps **1)** Put the "from" expression first: $(12x - 5) - (5x - 2y)$

2) Put in "1's": $1(12x - 5) - 1(5x - 2y)$

3) Use the distributive property: $12x - 5 - 5x + 2y$

4) Simplify or collect like terms: $7x + 2y - 5$

*****EXPONENTS DO NOT CHANGE IN ADDITION/SUBTRACTION*****

3.5

Multiplication: ANY Two Terms (like or unlike) can be multiplied.

Exponents of like variables are added in multiplication.

Monomial • Monomial: Use the example below

Example $(5x^2y)(-3xy^3z)$

Steps 1) Multiply the numerical coefficients: $(5)(-3) = -15$

2) Multiply the LIKE LETTER bases by ADDING their EXPONENTS: $(x^2)(x) = x^3$ and $(y)(y^3) = y^4$

3) Multiply the unlike letter bases by simply writing them down as part of the product: $-15(x^3)(y^4)(z)$

4) Write the answer as a product: $-15\,x^3y^4z$

Monomial • Polynomial: Use the distributive property. (See page 13)

Examples

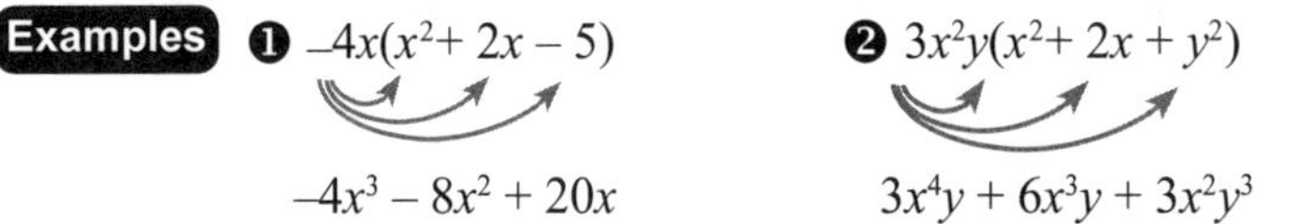

❶ $-4x(x^2 + 2x - 5)$ → $-4x^3 - 8x^2 + 20x$

❷ $3x^2y(x^2 + 2x + y^2)$ → $3x^4y + 6x^3y + 3x^2y^3$

Binomial • Binomial: Each term in the first binomial must be multiplied by each term in the 2nd binomial. Use the distributive property. We use the letters in FOIL to remind us to perform all the necessary multiplication steps. The letters in FOIL refer to the position of the terms in each parenthesis.

Multiply: **F**irst terms together,

Outside terms together,

Inside terms together, and

Last terms together. Then simplify. Use example below:

Example $(3x - 2)(x + 4)$

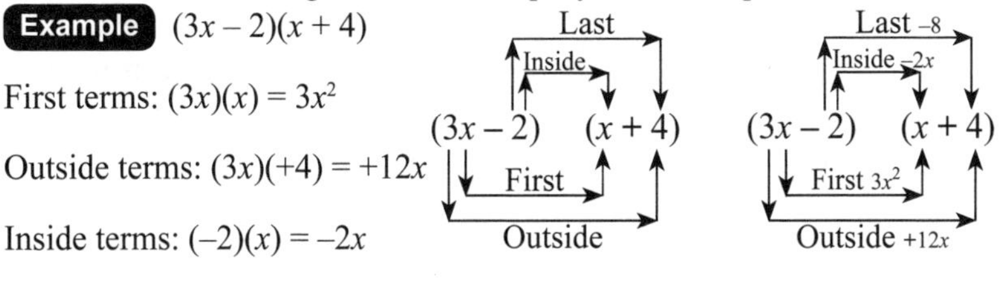

First terms: $(3x)(x) = 3x^2$

Outside terms: $(3x)(+4) = +12x$

Inside terms: $(-2)(x) = -2x$

Last terms: $(-2)(+4) = -8$

Collect like terms and simplify: $3x^2 + 12x - 2x - 8 = 3x^2 + 10x - 8$

Polynomial • Polynomial: Multiply each term in the first parenthesis by each term in the 2nd.

Example $(x - 2)(x^2 - 3x + 2) = x(x^2 - 3x + 2) - 2(x^2 - 3x + 2) =$
$x^3 - 3x^2 + 2x - 2x^2 + 6x - 4 = x^3 - 5x^2 + 8x - 4$

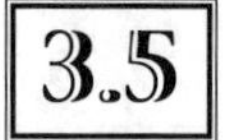

Binomial • Trinomial or larger multiplication problems.

Use the distributive property for multiplying any size polynomial by another. Multiply each term in the first polynomial by each term in the 2nd polynomial. Then combine like terms. Do this carefully so nothing gets lost. It is easy to overlook something.

Examples

❶ $(x+2)(x^2-3x+4) = x^3-3x^2+4x+2x^2-6x+8$
$= x^3-x^2-2x+8$

❷ $(x^3-2x+1)(x^2+5x-2) = x^5+5x^4-2x^3-2x^3-10x^2+4x+x^2+5x-2$
$= x^5+5x^4-4x^3-9x^2+9x-2$

My students prefer to use "vertical multiplication" when working with these bigger problems. It works very much like multiplying multiple digit numbers (like 345 • 7835) by each other.

Steps

1) Line the two polynomials up under each other. If one is smaller, put that one on the bottom.
2) Start at the right side and multiply each term of the top polynomial by the term furthest to the right on the bottom polynomial. Do this for each term in the bottom polynomial, moving right to left. Leave some extra space for missing exponents.
3) Add each column.

Examples

❶ Using vertical multiplication. $(x+2)(x^2-3x+4)$

$$
\begin{array}{rrrrl}
 & x^2 & -3x & +4 & \\
 & & x & +2 & \\
\hline
 & 2x^2 & -6x & +8 & \text{There are no missing exponents here.} \\
x^3 & -3x^2 & +4x & & \text{Line up with like terms.} \\
\hline
x^3 & -x^2 & -2x & +8 & \text{Add the columns.}
\end{array}
$$

❷ Using vertical multiplication. $(x^3-2x+1)(x^2+5x-2)$

$$
\begin{array}{rrrrrrl}
 & & x^3 & & -2x & +1 & \\
 & & & x^2 & +5x & -2 & \\
\hline
 & & -2x^3 & & +4x & -2 & \text{There is no } x^2 \text{ term so leave some extra space.} \\
 & +5x^4 & & -10x^2 & +5x & & \text{No } x^3 \text{ so leave space. Put the terms in columns with like terms.} \\
x^5 & & -2x^3 & +x^2 & & & \text{No } x^4 \text{ here, leave space and line up the terms carefully in columns.} \\
\hline
x^5 & +5x^4 & -4x^3 & -9x^2 & +9x & -2 & \text{Add each column.}
\end{array}
$$

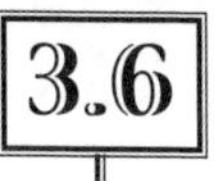

Solving Simple Equations

General Procedure for Solving Simple Equations: Do these steps in the order listed. Use the appropriate steps for a particular problem.

Steps

1) Remove parenthesis by multiplication (distributive property).

2) Collect like terms on each side of the equal sign.

3) Get all the terms containing the variable on one side of the equal sign and all number terms on the other by adding (add the opposite of what is already in the equation).

4) Separate the variable from its coefficient (multiply or divide by the coefficient).

5) Write the answer as $x = 5$ or SS = $\{5\}$ or SS = $\{x|x = 5\}$. Circle it.

6) CHECK the answer(s) in the ORIGINAL equation(s). Write the original down and show the substitution of your answer in that equation. Do the arithmetic on both sides of the equation. If the last step shows the two sides of the equation are equal to each other, smile! If not - redo the arithmetic in your check. If that is NOT OK, then DO THE PROBLEM AGAIN !!

Equations With One Variable: Isolate the variable (letter). Solve for the variable. (See page 65 for equations with two variables.)

Examples

❶ Solve for x: $x + 4 = 12$

Steps 1) Rearrange the number terms on one side and the terms containing x on the other: $x + 4 = 12$

2) Use subtraction to do this: $-4 \quad -4$

3) Answer: $x = 8$

❷ Solve for y: $5y - 5 = 15$

Steps 1) Get all the number terms on one side and the terms containing y on the other: $5y - 5 = 15$

2) Use addition to do this: $+5 \quad +5$

3) Isolate the term containing the variable: $5y = 20$

4) Solve for y: $\frac{5y}{5} = \frac{20}{5}$

5) Answer: $y = 4$

Note: Use multiplication or division to do this. x can only have "1" as a coefficient when you have solved the equation. [A negative variable ($-x$) is not an acceptable answer. Change the signs when needed by multiplying both sides of the equation by -1. (x) then becomes positive which is acceptable.]

Creating Equations

3.6

Decimal Equations: If the problem includes DECIMAL numbers for coefficients, multiply the entire equation by whichever power of 10 is necessary to remove the decimals and make all the numbers in the equation into whole numbers. Then solve as usual.

Example Solve for x: $0.5x + 2 = 17$

Steps
1) Remove the decimal and make: $10(0.5x + 2) = (17)(10)$
2) Whole numbers: $5x + 20 = 170$
3) Solve as usual for x: $5x = 150, x = 30$

Fractional Equation: If the equation contains FRACTIONS, it is best to "clear the fractions" by multiplying the entire equation by the least common denominator. This will remove the denominators and make the fractions into whole numbers. Then solve in the usual way.

Example Solve for x: $\frac{3}{4}x + 3 = \frac{15}{4}$

Steps
1) Multiply by 4 to remove the fractions: $(4)\left(\frac{3}{4}x + 3\right) = \left(\frac{15}{4}\right)(4)$
2) Isolate by subtracting: $3x + 12 = 15$
3) Divide by 3: $3x = 3$
4) **Answer:** $x = 1$

Equations With Letters for Coefficients

Example Solve for x: $ax + c = b$

Steps
1) Isolate the variable desired: $ax + c = b$
2) Divide both sides of the equation by the coefficient of the variable: $\underline{\quad -c \quad -c}$; $ax = b - c$
3) **Answer:** $x = \frac{b - c}{a}$

Rational Equations: These equations may have variables in the denominator of fractions that are in the equation. Remember that division by zero is undefined, so check to be sure that your solution checks in the problem. Extraneous solutions (those that do not check) must be marked accordingly. They are not in the solution set. The domain is often indicated to prevent the fraction from being undefined.

When solving, DO NOT "cancel" a variable to remove it from the fraction.

Examples on the next page

 **Rational Equations**

❶ $\frac{a-6}{a} - 1 = \frac{a+4}{a};\ a \neq 0$

$-\frac{a-6}{a} \qquad\qquad -\frac{a-6}{a}$ Rearrange so variables are on one side of =.

$-1 = \frac{a+4}{a} - \frac{a-6}{a}$

$-1 = \frac{(a+4)-(a-6)}{a}$ Combine fractions with like denominators.

$-1 = \frac{10}{a}$

$-a = 10;\quad a = -10$ Solve for a.

Check:

$\frac{-10-6}{-10} - 1 = \frac{-10+4}{-10}$

$\frac{-16}{-10} - 1 = \frac{-6}{-10}$

$\frac{6}{10} = \frac{6}{10}$ √

❷ $\frac{4}{x^2} = \frac{5}{x} - \frac{1}{x^2}$

$+\frac{1}{x^2} \qquad +\frac{1}{x^2}$

$\frac{5}{x^2} = \frac{5}{x}$ This can be treated as a proportion.

$5x = 5x^2$

$5x^1 - 5x = 0$

$5x(x-1) = 0$

$5x = 0;$ ~~$x = 0$~~ Extraneous solution. (Denominator cannot be zero.)

$x - 1 = 0; x = 1$ **Solution.**

Absolute Value Equations: Equations with terms contained in an absolute value symbol require careful handling. Use algebraic methods to isolate the absolute value symbol. Then, when solving, remember that the expression inside the absolute value symbol can be positive OR negative, so two answers are expected.

Creating Equations

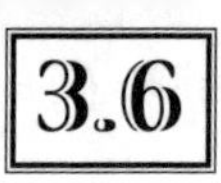

Examples **Absolute Value Equations**

❶ $|7x| - 3 = 67$

$\quad +3 \quad +3$ Isolate absolute value expression.

$|7x| = 70$

$7x = 70$ *or* $7x = -70$ Expression inside can be + or –.

$x = 10$; $x = -10$ $\quad SS = \{-10, 10\}$

❷ $2|x+8| = 86$

$$\frac{2|x+8|}{2} = \frac{86}{2}$$

$|x+8| = 43$

$x + 8 = 43$ *or* $x + 8 = -43$

$x = 35 : x = -51$

$SS = \{-51, 35\}$

<u>Exponential Equations</u>: Equations with a variable as an exponent are called exponential equations. At this level of math, they are solved by using a table of values or by graphing the equation and reading the answer from the graph.

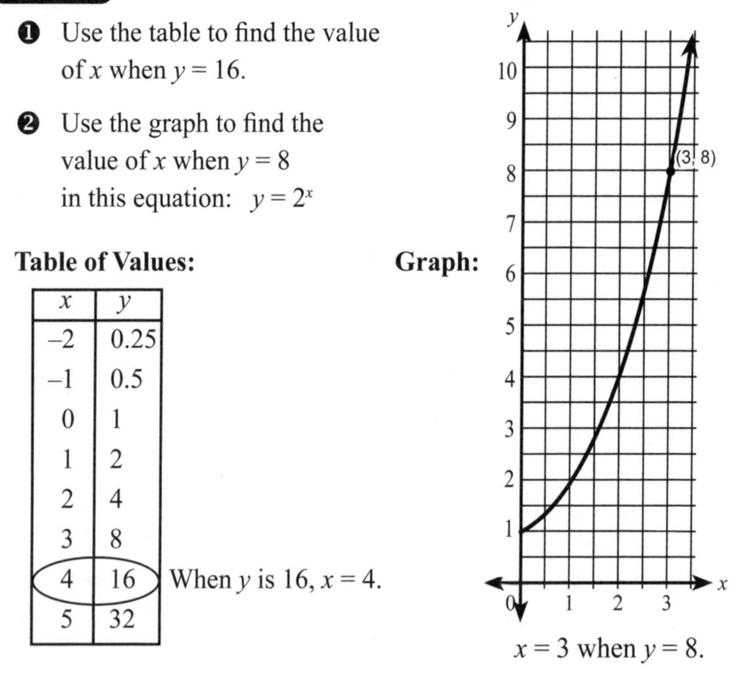

Examples

❶ Use the table to find the value of x when $y = 16$.

❷ Use the graph to find the value of x when $y = 8$ in this equation: $y = 2^x$

Table of Values:

x	y
–2	0.25
–1	0.5
0	1
1	2
2	4
3	8
4	16
5	32

When y is 16, $x = 4$.

Graph:

$x = 3$ when $y = 8$.

3.7 Creating Equations

Read all problems carefully.
It is easy to make a mistake in the setup.

WORD PROBLEMS USING VARIABLES

1. Represent the variables with letters.
2. Translate the words of the problem into algebraic phrases.
3. Solve for the variable.
4. Substitute the value of the variable in the "let" statement to find the other answers (when necessary).
5. Check your answer in the WORDS of the problem.
6. Make a conclusion - answer the question the problem asks - use words. Such as: "The number is 5." *or* "The rectangle is 6 by 7."
7. MAKE SURE your answer is correct for the domain (real numbers unless indicated otherwise). Sometimes an answer must be rejected.

Creating Equations

APPLICATIONS WITH 2 VARIABLES

Some problems require that variables are used to represent quantities in the equation that is created. When two variables are needed, two equations are created. In certain problems, two equations are solved separately, in others the two equations are solved as simultaneous equations or as a system of equations (both are solved together).

Examples

❶ Joe's Taxi service charges \$3.50 as a flat travel fee. He charges an additional \$0.45 per mile, x. Write an equation that represents his earnings, y.

$$y = 0.45x + 3.5$$

 Samuel has a jewelry business where his clients create their own bracelets. He charges \$25.00 for the bracelet and then \$6.00 for each bead. Write an equation that represents the cost, given x number of beads that are chosen.

$$C = 6x + 25$$

❸ The cost of operating Hannah's Biscotti Company is $1600 per week plus $.10 to make each biscotti cookie.

- Write a function, $C(b)$, to model the company's weekly costs for producing b biscotti. (See page 102)
  $C(b) = 1600 + 0.10b$
- What is the total weekly cost in dollars if the company produces 4,000 biscotti cookies?
  Total weekly cost
  $C(b) = 1600 + 0.10(4000) = \$2000$
- Hannah's company makes a gross profit of $0.60 for each biscotti cookie they sell. If they sold all 4000 biscotti they made, would they make money or lose money for the week (net profit)? How much?
  Find gross profit before calculating
  Gross Profit = $0.60b$
  Gross Profit = $0.60(4000) = \$2400$

  Net Profit = Gross Profit – Cost
  Net Profit = $2400 - 2000 = \$400$

  The company would make $400 profit for the week.

Number Problems: Use x for the number you know least about. Make the other parts of the "let" statement related to that one.

Example "One number is 21 less than twice the other number" Their sum is 54. Find both numbers

Steps:
1) Let x = one number and let $2x - 21$ = the other number.
2) This is the equation: $x + (2x - 21) = 54$
3) Then solve as usual:
$$3x - 21 = 54$$
$$+21 \quad +21$$
$$3x = 75$$
4) x is: $x = 25$
5) Substitute the value of x to find the other number: $2x - 21 \Rightarrow 2(25) - 21 = 29$
6) Answer: 25 and 29

Special Number Problems - sample "let statements"

Consecutive Integer problems: Let x = the first consecutive integer. $x + 1$ = the 2nd consecutive integer and $x + 2$ = the 3rd.

Positive or Negative Consecutive Integers: Let statement is still x, $x + 1$, $x + 2$, etc. Whether + or –, this "let" statement will work for all consecutive integer problems.

Odd or Even Consecutive integers: Let x = first odd integer, $x + 2$ = next consecutive odd integer. Even integers have the SAME let statement. Again, positive or negative odd/even integers have this same "let" statements. Remember zero is considered an even number.

Coin Problems: There are two things to consider in these problems. You must determine whether the problem is giving the NUMBER of coins (how many coins there are) or the VALUE of the coins (what the coins are worth in money). USUALLY THE "LET" STATEMENT tells about the NUMBER OF COINS, NOT THEIR VALUE. The value is often used in making the equation. When setting up the equation, if the problem uses dollars then use the decimal values (0.05, 0.10, 0.25) for the coins. If the problem is given in cents, use 5 for nickels, 10 for dimes, 25 for quarters. You may not need values: sometimes the problem asks only about how many coins there are and does not include any information about value.

Example Joe has $2.50. He has 7 more dimes than nickels. How many of each does he have?

Steps:

1) Set up the "Let" statement: Let x = number of nickels

2) Determine known values: $\therefore x + 7$ = number of dimes

3) Make equation and insert values of monies: $.05x + .10(x + 7) = 2.50$

4) Multiply out (distributive property): $.05x + .10x + .70 = 2.50$

5) Isolate x: $.15x + .70 = 2.50$

$$\underline{\quad -.70 \quad -.70}$$

6) Solve for x: $\frac{.15x}{.15} = \frac{1.80}{.15}$

7) Answer: $x = 12$ nickels

8) Plug answer back into "Let" statement: $x + 7 = 19$ dimes

9) Answer: He has 19 dimes and 12 nickels

Note: $\therefore$ means "therefore"

Ratio Problems: Word problems sometimes involve ratios between numbers or items. Use the ratio information given in the problem to make the "let" statement. Then use the "let" statement to make an equation for the solution.

Example Find the measure of each angle of a triangle whose angles have a ratio of 3:6:9. (Use information you already know about triangles to form the equation: the sum of the angles in a triangle is 180°.)

Steps:

1) Set up the "Let" statement: Let $3x$ = one angle, $6x$ = the 2nd angle, and $9x$ = the 3rd angle.

2) Make the equation: $3x + 6x + 9x = 180$

3) Solve for x: $18x = 180°$ *or* $x = 10$

4) Plug the answer in the "Let" statement to find the three angles: $(10) = 30$; $6(10) = 60$; and $9(10) = 90$

5) Answer: The three angles measure 30°, 60°, and 90°.

Creating Equations

3.7

Age and Mixture Problems: Make a chart to show the information in the problem. This can be used as the "let" statement (or you can use it to make your written "let" statement). Make columns for the information given: Use the chart below to make an equation.

Examples

❶ Sue is 5 years older than Ann. In 6 years, Sue's age will be 11 years less than twice Ann's age then. How old is each person now?

Let Statement

NAME	AGE NOW	AGE IN 6 YEARS
Ann	x	$x + 6$
Sue	$x + 5$	$(x + 5) + 6$

Steps:

1) Analyze the phrases mathematically and make a chart:
   In 6 years (use "age in 6 years" column) Sue's age $[(x + 5) + 6]$ will be (=) 11 years less than (–11) twice (2) Ann's age then $(x + 6)$.
2) Make an equation: $(x + 5) + 6 = 2(x + 6) - 11$
3) Use distributive property: $x + 11 = 2x + 12 - 11$
4) Solve for x: $x + 11 = 2x + 1$
   $-x \quad -1 \quad -x \quad -1$
5) Ann's current age: $10 = x$
6) Plug x back in to find Sue's age: $x + 5 = (10) + 5$, so Sue's age = 15
7) **Answer:** Ann is now 10 and Sue is now 15.
8) Check: In 6 years Ann will be 16, Sue will be 21 (21 is eleven less than twice 16).

❷ The 9th grade students in Public School #4 are selling tickets to the class play. Tickets purchased by students will cost \$3.00 each and tickets sold to the public will cost \$5.00 each. They sell 400 tickets but do not keep track of whether they are purchased by students or by the public. They have \$1500 after selling the tickets. How many of each type of ticket did they sell?

Type of Ticket	Tickets Sold	Price per Ticket	Money Earned
Student	x	\$3.00	\$3.00x
Public	$400 - x$	\$5.00	\$5.00$(400 - x)$
Total	400		\$1500

$$3x + 5(400 - x) = 1500$$
$$3x + 2000 - 5x = 1500$$
$$-2000 \qquad -2000$$
$$-2x = -500$$
$$x = 250$$
$$400 - x = 400 - 250 = 150$$

Answer: 250 tickets were sold to students and 150 were sold to the public.

Geometry Word Problems: Draw a diagram to demonstrate the problem. Label it with the information for the "let" statement. Substitute what is given in the formula.

(Example on top of next page.)

Example A garden has a perimeter of 48 feet. If one side of the garden is 2 feet shorter than the other side, what are the dimensions of the garden?

Diagram:

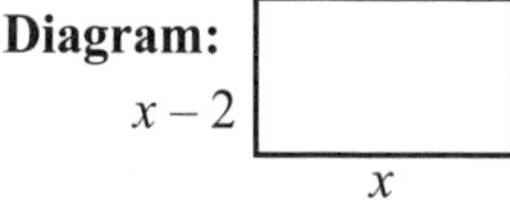

or use a "let" statement

Let x = the length

Let $x - 2$ = the width

Formula: $P = 2(l + w)$

$$48 = 2(x + x - 2)$$
$$48 = 2(2x - 2)$$
$$48 = 4x - 4$$
$$+4 \qquad +4$$
$$52 = 4x$$
$$\frac{52}{4} = \frac{4x}{4}$$
$$13 = x$$
$$x - 2 \Rightarrow 13 - 2 = 11$$

Conclusion: The garden is 13 ft by 11 ft.

PERCENTS – INCREASE, DECREASE, AND DISCOUNT

Discount: Stores often give a % off as a discount. Find the sale price by using this: *Original selling price – Discount amount = Discount sale price*

Note: The dollar sign is usually omitted in the formula but must be put back in the answer when the answer is in dollars.

Examples

❶ Jerome bought a pair of sneakers at a 25% discount sale. He paid $60.00. What was the original price of the sneakers?

Original Price = x

Discount amount = $0.25x$ (Change the % to a decimal.)

Sale Price = $60

Substitute in the formula:

$$x - 0.25x = 60$$
$$0.75x = 60$$
$$x = 80$$ The original price was $80.

❷ Find the cost of a $200 bike that is on sale at 15% off.

Original Price = $200, Discount Amount = (0.15)($200),

Discount sale price = x

$200 - (0.15)(200) = x$ Substitute

$200 - 30 = x$ Solve for x

$x = 170$ The discount sale price of the bike is $170.

❸ What is the discount on a skateboard that was originally sold for $150 and is now on sale for $120?

Original Price = $150, Discount Amount = ($x$)($150),

Discount sale price = $120

$150 - 150x = 120$ Substitute

$-150x = 120 - 150$

$x = \frac{-30}{-150}$ Solve

$x = 0.2$ (Change to a % by multiplying by 100%)

The discount is 20% off.

Decrease: Another way to say this is that the sale price is a *decrease of 20%* from the original price. This is called a percent of decrease.

(Examples continued from previous page.)

❹ My pool was 72°F on Monday. On Friday it was 80°F. What is the **percent of increase**, to the nearest tenth, of the temperature in the pool? This is not a money problem, but it can be handled in a similar way. Since the temperature is increasing, use + the change for the 2nd term.

Original + change = final result
Original temperature (72) + % change (x) = Final temperature (80)
$72 + (x)(72) = 80$
$72x = 8$
$x = 0.1111...$ Change to % = 11.111...%
The pool temperature increased by about 11.1%

__**Similarity**__**:** 2 polygons are similar if they have corresponding sides that are proportional and corresponding angles that are congruent. The order in which the letters identifying the vertices in one polygon are written to match the order of the corresponding vertices of the 2nd polygon. To find the lengths of sides of similar polygons, we use proportions.

__**Solving Geometry Problems with Algebraic Fractions and Proportions**__

Problems using proportions:

1. Draw diagrams of both similar figures and label the known sides.
2. Make two equal ratios using information from one polygon for the numerators of both fractions and corresponding information from the other polygon as the denominators of both fractions.
3. Multiply the numerator of one fraction by the denominator of the other. Then multiply the denominator of the first fraction by the numerator of the second fraction. This is commonly called "cross multiply". Solve.

Example ΔABC is similar to ΔRST. If $AB = 12$ and $BC = 6$, find the lengths of RS and ST if RS is 2 units more than ST.

Steps 1)

C, 6, B, A, 12 T, x, S, R, $x + 2$

2) Match up similar sides: $\overline{AB} \sim \overline{RS}$, $\overline{BC} \sim \overline{ST}$

3) Set up the proportion: $\frac{6}{x} = \frac{12}{x+2}$

$6(x + 2) = 12x$

4) Solve for x:

$6x + 12 = 12x$
$-6x \quad\quad -6x$
$12 = 6x$
$x = 2$

5) Conclusion: $ST = 2$, *and* $RS = 4$

Scale Drawings: A scale drawing is a reduction or an enlargement of a real object. Architectural drawings, models, and maps are some examples of scale drawings.

Scale: The RATIO in the drawing. Use the ratio that is the scale of the drawing in a proportion to find information about the drawing or the real object.

General Proportion: $Scale\ (a\ fraction) = \dfrac{Drawing}{actual\ or\ real\ object}$

Example The scale of a map is 1 cm to 5 km. The distance on the map from Santa Fe to Johnstown is 4.5 cm. What is the actual distance between these two cities?

The SCALE is 1 cm : 5 km., the distance in the map or "drawing" is 4.5 cm, and we don't know the real object distance.

Steps 1) Use a variable for the real object distance and make a proportion: $\dfrac{1cm}{5km} = \dfrac{4.5cm}{x}$

2) Cross multiply: $1cm \bullet x = (4.5cm)(5km)$

3) Divide by 1 cm to simplify: (In this step, the "*cm*" cancels.) $\dfrac{1cm \bullet x}{1cm} = \dfrac{(4.5\not{cm})(5km)}{1\not{cm}}$

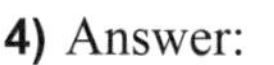

4) Answer: $x = 22.5\ km$

Note: Include the units of measure. Some will cancel while the work is being performed, giving the correct units for the final answer. The actual distance from Santa Fe to Johnstown is 22.5 km.

Exponential Equations: An exponential equation has a variable in the exponent. For now, these problems can be solved by making a table or a graph.

Example Nathan owns 4 rabbits. He expects the number of rabbits to double every year. After how many years will he have 64 rabbits?

Let x = number of years
Let y = number of rabbits

Write an equation to model this situation.

Answer: $y = 4(2)^x$

Use a table or a graph to solve the equation. [The use of a grid is optional.]

Answer: 4 years and appropriate work is shown.

(See also Unit 4)

x	y
0	4
1	8
2	16
3	32
4	64
5	128

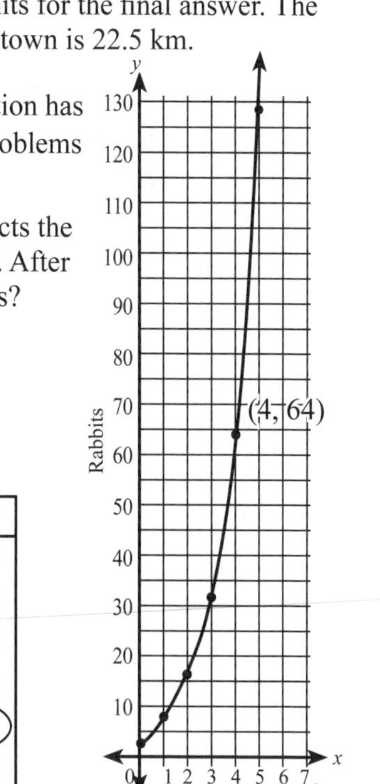

Creating Equations

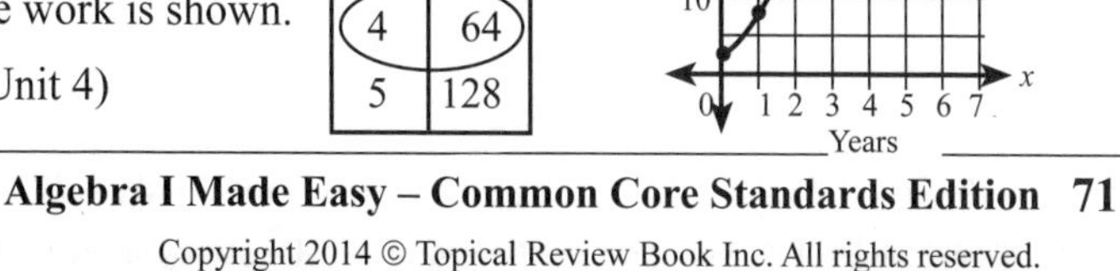

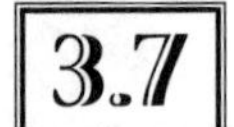

3.7

Solving Equations with 2 Variables Separately

Steps **1)** Solve the first equation as usual.

2) Substitute the result from the first equation into the second one.

3) Make sure to answer the question(s) completely.

Examples

❶ Julio rides his bike to school and his rate of speed is 8 miles per hour. He gets to school in 20 minutes. Mike jogs to school the same distance and it takes him 30 minutes. How far does each boy go to get to school and what is Mike's rate of speed?

Let x = Julio's distance
Let y = Mike's rate of speed
Formula: $d = rt$

Julio	Mike
$r = 8$ mph, $t = \frac{20}{60}$ or $\frac{1}{3}$ hr	$d = 2\frac{2}{3}$ mi, $t = \frac{30}{60}$ or $\frac{1}{2}$hr
$8 \cdot \frac{1}{3} = x$	$y \cdot \frac{1}{2} = x;\ y \cdot \frac{1}{2} = 2\frac{2}{3}$
$x = 2\frac{2}{3}$ miles	$y = 2\frac{2}{3} \cdot 2$
	$y = 5\frac{1}{3}$ mph

Conclusion: The distance traveled by each boy is $2\frac{2}{3}$ miles. Mike's rate of speed is $5\frac{1}{3}$ miles per hour.

❷ Tammy wants to make the most money she can while maintaining a 90 average. Currently she makes $9.00 per hour at her job. She knows she can maintain a 100 average if she does not work at all. She has observed that for every two hours of work, her average goes down one point. How many hours can she work and how much money can she make each week while still maintaining a 90 average?

Solution: This is a two part problem. Write an equation to represent the maintaining a 90 average. (Notice that her average goes down only one point for two hours of work.) Make another equation to determine the amount of money she can earn working that number of hours.

Hours Tammy can work, x:

$$100 - .5x = 90$$
$$\underline{-90 + .5x \quad -90 + .5x}$$
$$10 = .5x$$
$$x = 20$$

Money she can earn, y:

$$y = 9x$$
$$y = 9(20)$$
$$y = 180$$

Conclusion: Tammy can work 20 hours and earn $180 while maintaining a 90 average.

3.7

Solving Equations with 2 Variables Simultaneously

Steps 1) Identify each unknown quantity and represent each one with a different variable in a let statement. READ CAREFULLY. Make the let statement accurate.

2) Translate the verbal sentences into *two* equations.

3) Solve as a system of equations. Usually these problems are solved algebraically but follow directions - you might be directed to solve them graphically.

4) Check the answers in the words of the problem.

Examples

❶ Together Evan and Denise have 28 books. If Denise has four more than Evan, how many books does each person have?

Let x = the number of books Evan has

Let y = the number of books Denise has

Steps

1) Set up equation (A): $x + y = 28$

2) Set up equation (B): $y = x + 4$

3) Use substitution (See page 79):

$$x + (x + 4) = 28$$
$$2x + 4 = 28$$
$$-4 \quad -4$$

4) Solve for x:

$$2x = 24$$
$$x = 12$$
$$x + y = 28$$

5) Substitute in original: $12 + y = 28$

6) Solve for y:

$$-12 \quad -12$$
$$y = 16$$

Answer: Evan has 12 books and Denise has 16 books.

❷ The length of a rectangular fence is 4 times the width. The perimeter is 200 feet. Find the length and the width of the fence.

Steps

1) Perimeter formula: $P = 2l + 2w$

2) Let Statement: l = length, w = width

3) Based on the problem analysis, write l in terms of w: $l = 4w$

4) Substitute in the formula: $200 = 2(4w) + 2w$

5) Solve for w:

$$200 = 10w$$
$$w = 20$$

6) *Substitute* 20 for w *to find the length*: $l = 4(20) = 80$

Answer: The length of the fence is 80 feet and the width is 20 feet.

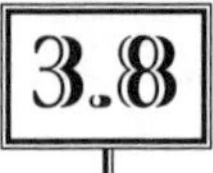

Graphing Linear Equations

Linear Equation: An equation that is graphed on a coordinate graph which forms a straight line. The line represents all the values of x and y that will make the equation of that line true. The coordinates of any point on the line can be substituted for x and y, in the correct order, in the equation of the line and they will check. They are its solutions.

Collinear Points: Two or more points that are on the same line. To find out if two points are collinear, substitute the (x, y) values of one in the equation of the line. Then substitute the coordinates of the other in the same equation. If both sets satisfy the same equation, then they are collinear.

Independent Variable: This is the variable whose value is shown on the horizontal axis, (x).

Dependent Variable: The variable whose value is shown on the vertical axis, (y).

Slope: The ratio of the change in y to the change in x of a graphed line – it tells the "steepness" of the graph. It is usually written as a fraction. The symbol for slope is "m" and that is used in formulas but it is not acceptable as part of the answer. The word "slope" should be used.

Slope Intercept Form: $y = mx + b$ is the Slope Intercept Form of a linear equation, where "m" represents the slope and b represents the y value of the y-intercept.

Working with Slope:

Using Slope: Starting at a point on the line, the slope shows how to locate the next point on a graph line. The numerator (top) of the slope fraction indicates vertical movement on the graph, the denominator (bottom) number indicates horizontal movement. If slope is shown as an integer rather than a fraction, it can be made into a fraction by putting "1" in the bottom of the fraction to serve as the denominator.

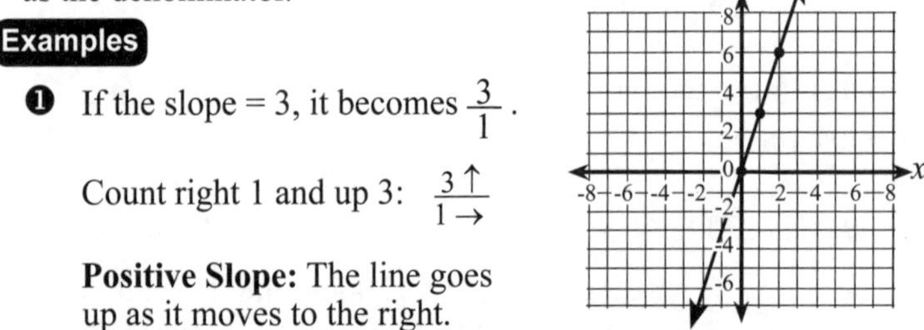

Examples

1. If the slope = 3, it becomes $\frac{3}{1}$.

   Count right 1 and up 3: $\frac{3 \uparrow}{1 \rightarrow}$

   Positive Slope: The line goes up as it moves to the right.

❷ If the slope = –5, it becomes $\frac{-5}{1}$.

Count right 1 and down 5: $\frac{-5\downarrow}{1\rightarrow}$

Negative Slope: The line goes down as it goes to the right.

Reminder: $\frac{-5}{1} = -\frac{5}{1} = \frac{5}{-1}$. For slope, it is easiest to use $\frac{-5}{1}$.

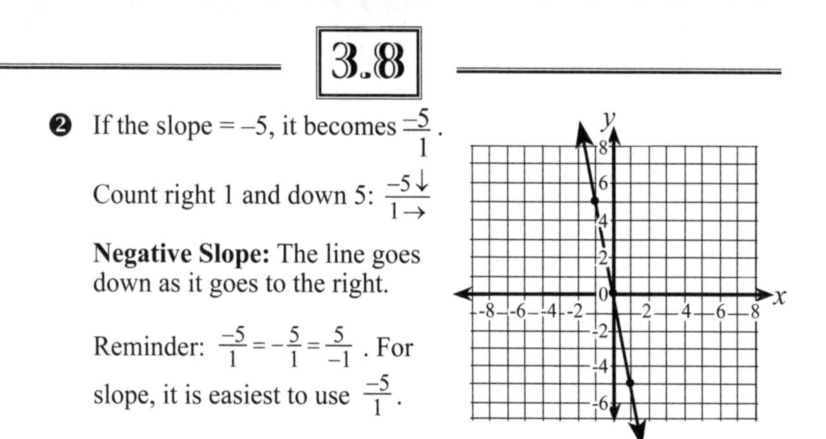

Need help remembering slope?? The numerator, or upper number of the slope fraction, indicates the up or down (vertical) count for the slope. The denominator or lower number indicates the count to the right or left (horizontal count). The positions of the labels on the graph axes match the positions of the numbers in the slope fraction. The label for the y-axis is at the top of the graph, and the vertical count for the slope is in the top (numerator) of the fraction. The label for the x-axis, the horizontal axis, is located in a lower position on the graph, and the horizontal count for the slope is in the lower position of the fraction (the denominator).

<u>**Finding The Slope From 2 points**</u>: **Slope formula:** $m = \frac{\Delta y}{\Delta x} = \frac{y_2 - y_1}{x_2 - x_1}$

Δ is the Greek letter delta and it stands for "change in".

To find the slope of a line if you know two points on the line, name the points 1 and 2. Then substitute the y values for 1 and 2 in the formula and the x values for 1 and 2 in the formula. These must be taken in the same order.

Find the slope of a line passing through (3, 2) and (–4, 5).

Steps

1) Point 1 is (3, 2), Point 2 is (–4, 5).

2) Determine x and y values: $x_1 = 3$ $\quad x_2 = -4$
$y_1 = 2$ $\quad y_2 = 5$

3) Plug Points into formula: $m = \frac{5-2}{-4-3} \Rightarrow$ slope of the line $= \frac{3}{-7}$

[Remember: $\frac{-3}{7} = -\frac{3}{7} = \frac{3}{-7}$]

3.8

Finding The Slope From an Equation: Solve the equation for y in terms of x. Put it in the form $y = mx + b$. The coefficient of the x variable (after the equation is solved for y in terms of x) is the slope of the line of that equation.

Example What is the slope for the line $2y - 4x = 6$?

Steps

1) Isolate and solve for y: $2y - 4x = 6$

$2y = 4x + 6$

2) Use $y = mx + b$: $y = 2x + 3$

3) Determine slope (m): $m = 2$ or $2/1$

(The slope is used as a fraction. Use 1 for the denominator if m is not already a fraction.)

Finding The Slope Using A Straight Line Already Graphed: Locate a point and then count the horizontal and vertical distance to the next point to find the slope ratio.

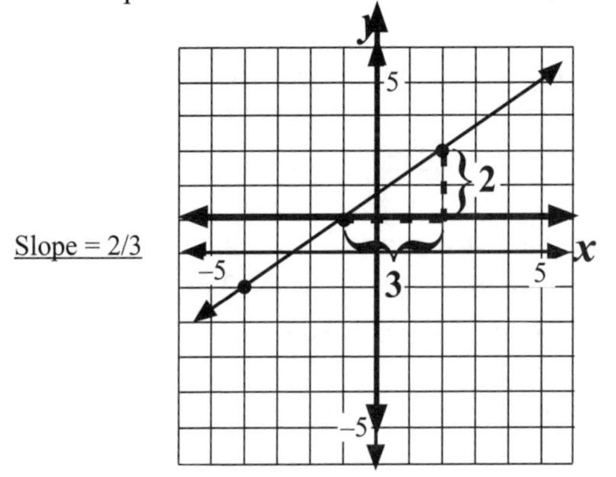

Special Slopes - Memorize these

slope = 0

If y = *a number*, and the equation has no x, the graph is a **horizontal** line and parallel to the x-axis. It crosses the y-axis at the number given in the problem.
The slope = 0.

A line parallel to the x-axis will never cross the x-axis. x doesn't even appear in the equation. The equation will always be $y =$ ____ (blank is filled with a number, the appropriate value of y.)

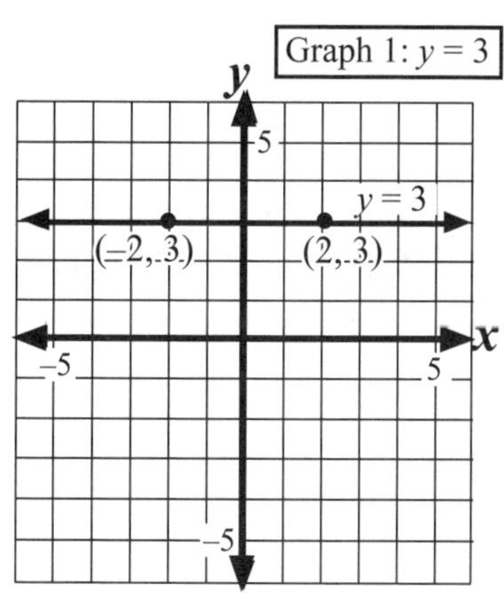

Example

Write an equation through the point (2, 3) that is parallel to the x-axis. The equation is $y = 3$.

Undefined Slope

If x = *a number*, and the equation has no y, the graph is a vertical line and parallel to the y-axis. It crosses the x-axis at the number given and **its slope is undefined**. Sometimes we say it has no slope.

A line parallel to the y-axis will never crosses the y-axis, there is no y in its equation. The equation will always be $x =$ ____ (blank is filled with a number, the appropriate value of x.)

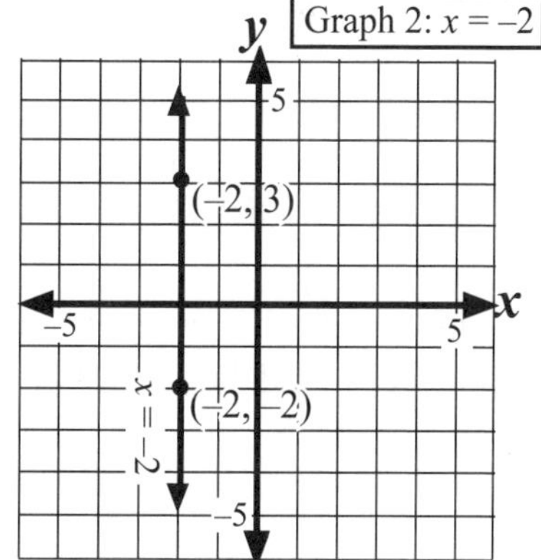

Example

Write an equation through the point (–2, 4) that is parallel to the y-axis. The equation is $x = –2$.

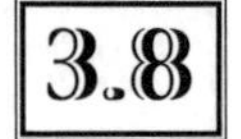

<u>Y-Intercept</u>: The point at which the graph line crosses the y-axis. Coordinates are always $(0, y)$. The x value is always 0 and y is the number on the y-axis where the line crosses.

- Solve the equation for y in terms of x and put it in the form $y = mx + b$. In this form, with y isolated, b is the y value of the y-intercept. As shown in the previous example, $y = 2x + 3$, the y-intercept is $(0, 3)$.
- The y-intercept can also be found by substituting "0" for x in the equation and then solving for y. The value of y when $x = 0$ is the y coordinate of the intercept.

Example What is the y-intercept for the line $2y = 4x + 2$?

Steps

1) Insert "0" for x: $2y = 4(0) + 2$

2) Solve for y: $2y = 2$; $y = 1$

3) Answer: y-intercept is $(0, 1)$

<u>Collinear Points</u>: Points that are on the same line. If the slope is the same between the first and second points as it is between the second and third (or first and third), then the points are on the same line. Use the slope formula to test this.

Example Are (3, 5), (6, 8), and (–4, –8) collinear points?

Solution: Call (3, 5) point 1, (6, 8) point 2, and (–4, –8) point 3. First, find the slope between points 1 and 2. Then use the same formula for points 2 and 3.

Slope of points 1 and 2

$$m = \frac{8-5}{6-3} = \frac{3}{3} = 1$$

Slope between points 2 and 3

$$m = \frac{-8-8}{-4-6} = \frac{-16}{-10} = \frac{8}{5}$$

Since the slope between the first two points is 1, and the slope between the second and third points is $\frac{8}{5}$, thus not the same, these three points are not on the same line.

Parallel and Perpendicular Lines and Slope: Use the slope formula or the slope intercept form of an equation ($y = mx + b$) to determine if two lines are parallel, perpendicular, or neither.

- **If two lines have the same slope**, the lines are parallel or they may be the same line.

Example Are $y = 3x - 4$ and $y = 3x + 7$ parallel lines?
They both have a slope of 3, so yes, they are parallel lines.

Note: If two equations have the same slope and the same y-intercept, they become the same line when they are graphed.

- **If two lines are parallel**, they have the same slope.

Example Write an equation of a line parallel to the line $y = -3x + 2$.

The slope of this line is –3 so the new equation will have the same slope but a different y-intercept.

It could be $y = -3x + 5$ *or* $y = -3x - 2$, *or* just $y = -3x$ (the y-intercept is zero).

- **If two lines have slopes that are negative reciprocals of each other**, the lines are perpendicular.

Example Are these two lines perpendicular? $y = 3x - 4$ and $y = \frac{-1}{3}x + 12$.
Since the slopes are 3 and –1/3 which are negative reciprocals of each other, the lines are perpendicular.

- **If two lines are perpendicular**, their slopes are negative reciprocals of each other.

Example Name two points that would be on a line that is perpendicular to the line $y = -2x$.

To solve this, we must find two points that will work in the slope formula to make a slope of $+\frac{1}{2}$. The line's equation will be $y = \frac{1}{2}x$ or $y = \frac{1}{2}x$ + some constant. An answer might be (6, 3) and (12, 6).

3.9

SYSTEMS OF EQUATIONS

(Simultaneous Equations)

Linear Equation: A first degree equation which will form a straight line when its solution set is graphed. It can contain one or two variables.

Linear Pairs: Two first degree equations that are solved together. Linear pairs can be solved algebraically or graphically. Their solution set as a pair of equations is the ordered pair whose values make both equations true. Sometimes a linear pair has no solution.

Solution: The values of x and y that make both equations true. An algebraic solution is considered exact while a graphic solution is approximate.

Graphing: Each equation will be graphed on the same coordinate axis (same graph). The solution is the point(s) of intersection of the two graphs. (See also page 81) If the lines are parallel there is no solution. If the two lines are concurrent (in the same place) the solution is all the real numbers.

Algebraic Solution: Two methods are used -- substitution or addition.

1. **Substitution:** One equation is manipulated so that x or y is isolated, then the resulting representation of that variable is substituted in the 2nd original equation. The remaining variable is solved for, then that answer is substituted in either original equation to find the second variable.

Example Solve both equations $x - y = 1$, $x - 2y = 3$

Steps:

1) Solve the first equation ($x - y = 1$) for "x": $x = y + 1$
2) In the second equation ($x - 2y = 3$), substitute ($y + 1$) for x: Solve for y:
$$(y + 1) - 2y = 3$$
$$-y + 1 = 3$$
$$-y = 2$$
$$y = -2$$
3) Go back to an original equation: $x - y = 1$
4) Substitute -2 for y: $x - (-2) = 1$
5) Solve for x: $x + 2 = 1$
6) Indicate both answers: $x = -1$ and $y = -2$
7) Checking both answers in both original equations is the final step:
$$x - 2y = 3$$
$$-1 - 2(-2) = 3$$
$$-1 + 4 = 3$$
$$3 = 3$$

Note: This method is recommended only when the coefficient of x or y is 1. Coefficients other than 1 can result in fractional substitutions which must be "cleared" or they rapidly become unmanageable.

Algebraic Solution (continued)

2. **Addition:** If two equations have the same variable with opposite coefficients, we can add the equations together and eliminate that variable. Sometimes it is necessary to multiply one equation (or both) to make equivalent equations that can be used in this method.

Steps

1) Arrange both equations using algebraic methods so the variables are underneath each other in position.
2) The goal is to eliminate one variable by adding the two equations together. Examine the variables and their coefficients. Find the least common multiple of the coefficients of either both x's or both y's.
3) Multiply each equation by a positive or negative number so the coefficients of the variable chosen are equal and opposite in sign.
4) Add the two equations together. One variable will disappear.
5) Solve for the variable that is visible.
6) Choose one of the *original* equations and substitute the value for the known variable to find the other variable.
7) *Check in both original* equations.

Example Solve the following equations for x and y:
(A) $4x + 6y = 64$ (B) $2x - 3y = -28$

Steps

1) 6 is a common multiple for 6 and 3 so leave (A) as is and multiply (B) by 2: $2(2x - 3y = -28) : 4x - 6y = -56$
2) Add (A) and (B) in order to isolate x:
Equation (A): $4x + 6y = 64$
NEW Equation (B) from step 1: $\underline{4x - 6y = -56}$
$8x = 8\ ;\ x = 1$
3) Insert the answer back into either (A) or (B): $4(1) + 6y = 64$
4) Solve for y: $6y = 60\ ;\ y = 10$
5) Check in both *original* equations: $4(1) + 6(10) = 64$ $\quad 2(1) - 3(10) = -28$
$64 = 64\surd$ $\quad -28 = -28\surd$

Note: Experience will help you decide which variable to work with in the addition method. Looking for variables which already have opposite signs allows you to avoid multiplying by a negative number with its associated opportunities for error. Using small multiples is helpful, too. You wouldn't want to use a common multiple for 11 and 14 if you could use the other variable and have a common multiple of 2 and 5.

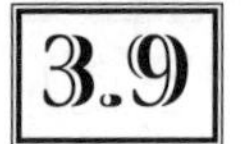

3.9

Graphing Systems of Linear Equations: Two or more equations are graphed on the same coordinate plane (grid).

- Graph EACH equation separately but put both on one coordinate graph. Be ACCURATE.
- Label each line as you graph it.
- The point where they intersect is the approximate solution set of the system of equations.
- Label the point of intersection on the graph. This point is the solution set.
- Check both the x and y values of the solution in both original equations. The x and y values of the point of intersection must satisfy both equations.

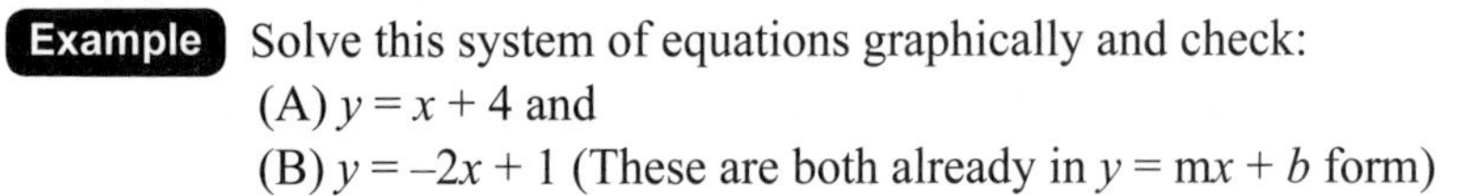

Example Solve this system of equations graphically and check:
(A) $y = x + 4$ and
(B) $y = -2x + 1$ (These are both already in $y = mx + b$ form)

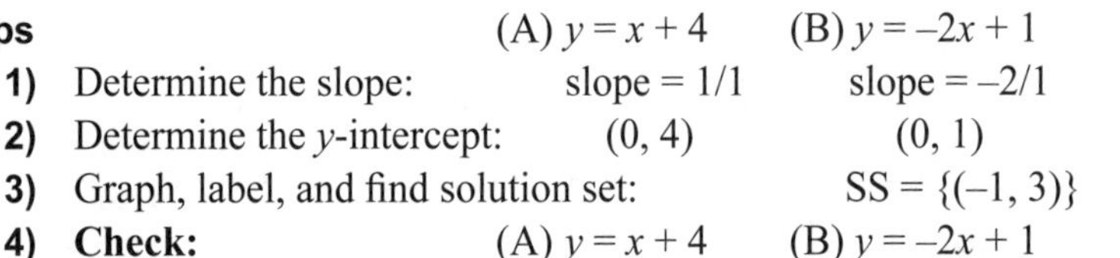

Steps	(A) $y = x + 4$	(B) $y = -2x + 1$
1) Determine the slope:	slope = 1/1	slope = –2/1
2) Determine the y-intercept:	(0, 4)	(0, 1)
3) Graph, label, and find solution set:		SS = {(–1, 3)}
4) Check:	(A) $y = x + 4$	(B) $y = -2x + 1$
	$3 = -1 + 4$	$3 = -2(-1) + 1$
	$3 = 3$ √	$3 = 2 + 1 \Rightarrow 3 = 3$ √

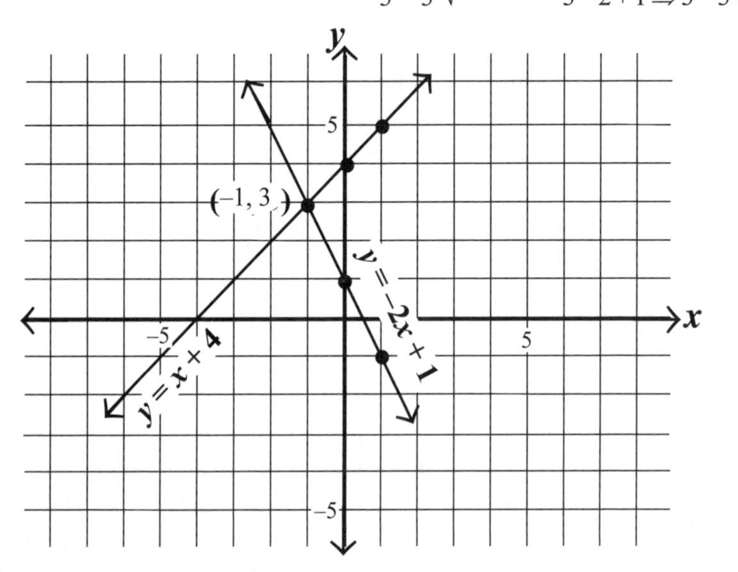

Solution: (–1, 3) is the point at which $y = x + 4$ and $y = -2x + 1$ are equal.

Graphing Systems of Non-Linear Equations: Make a table of values for both equations and find the points where they are alike.

OR Graph both equations and identify the point where they intersect. It is at this point that the equations are equal. The x and y values of either method should check in both equations.

Examples

❶ $y = x^2$ This pair of equations is called a quadratic linear pair.
$y = -2x$

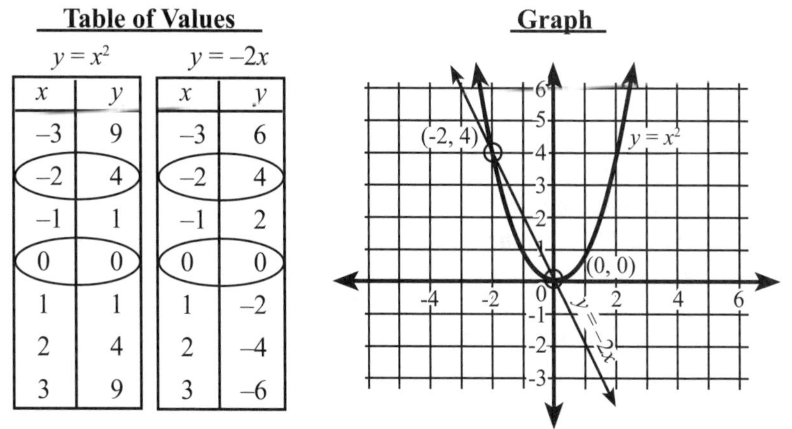

Table of Values

$y = x^2$

x	y
–3	9
–2	4
–1	1
0	0
1	1
2	4
3	9

$y = -2x$

x	y
–3	6
–2	4
–1	2
0	0
1	–2
2	–4
3	–6

Graph

The values (–2,4) and (0,0) satisfy both equations.

❷ Solve this pair of equations graphically. This is a pair of exponential equations.

$y = 2^x - 2$
$y = 0.5^x - 2$

Solution: (0,–1)

Graph:

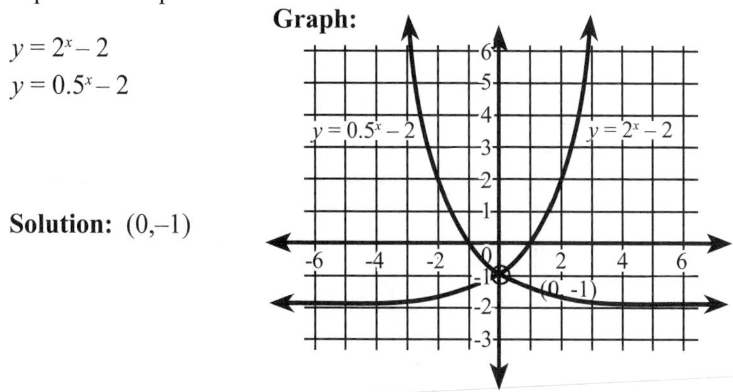

Calculations of the approximate point of intersection may be done on the calculator when reading the graph for non-integer answers.

❸ Solve this pair of absolute value equations using the table of values.

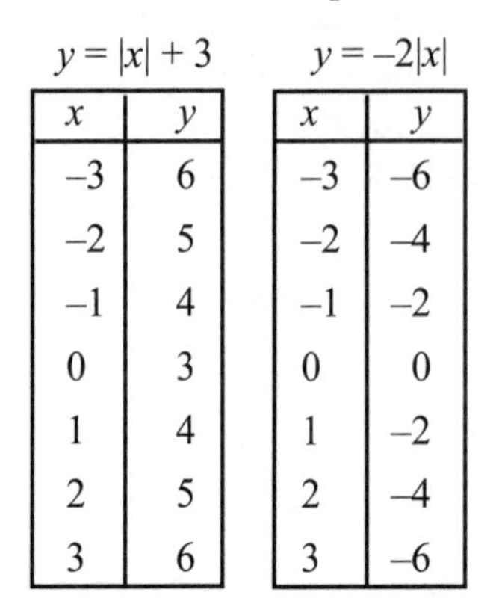

$y = |x| + 3$

x	y
–3	6
–2	5
–1	4
0	3
1	4
2	5
3	6

$y = -2|x|$

x	y
–3	–6
–2	–4
–1	–2
0	0
1	–2
2	–4
3	–6

Solution: This pair of equations does not have a solution. Extending the table of values will make both sides of the first equation, $y = |x| + 3$, continue upward in a y direction, and the y values in the 2nd equation, $y = -2|x|$, will continue downward. There is no solution. These can be quickly graphed on the calculator to verify your conclusion. The graphs do not intersect as shown below.

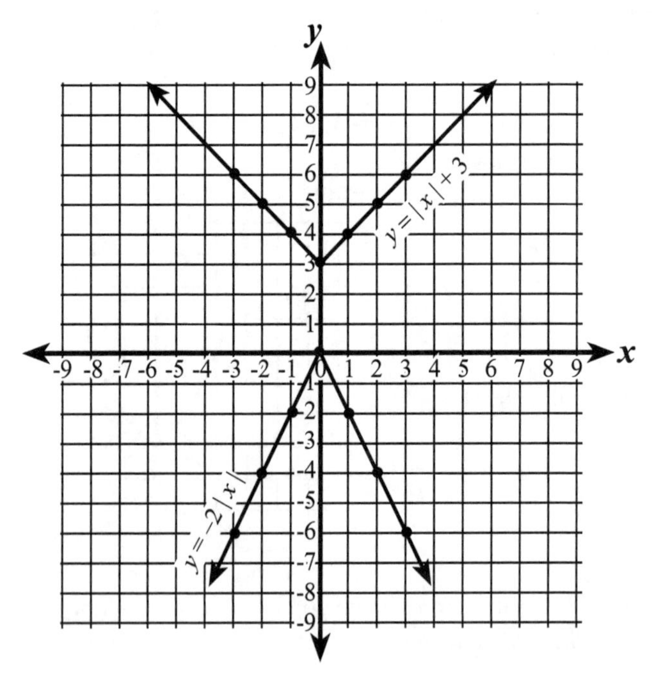

Solving Inequalities

Inequalities can be solved algebraically or geographically. The SOLUTION SET for an inequality is a group of numbers rather than a single number. We solve simple inequalities using the same algebraic procedures that we use for equations EXCEPT when we multiply or divide by a negative number. When multiplying or dividing by a negative number, the DIRECTION of the inequality sign is reversed, < becomes >, and > becomes <.

The solution set for inequalities is not always a finite set like the answers are for equations. Write the solution set as SS = $\{x|\, x > 5\}$ for example. If the domain is whole numbers, then write SS = {6, 7, 8, ...} for example. If the solution set is between certain numbers, write $\{x|\, 3 < x < 10\}$. If the answer is integers greater than 3 and less than 10, write SS = {4, 5, 6, 7, 8, 9}. If the domain is integers or whole numbers, it may be necessary to round a fractional or decimal answer up or down so it solves the equation and is in the domain. In word problems, it is sometimes necessary to answer with a whole number answer - so read the problem carefully.

ALGEBRAIC SOLUTION

Examples

❶
$$\begin{aligned} 3x - 4 &\le -12 \\ +4 \quad &\;\; +4 \\ \hline 3x &\le -8 \\ \frac{3x}{3} &\le \frac{-8}{3} \\ x &\le -2\tfrac{2}{3} \end{aligned}$$

❷
$$\begin{aligned} -2x + 3 &> 5 \\ -3 \quad &\;\; -3 \\ \hline -2x &> 2 \\ \frac{-2x}{-2} &< \frac{2}{-2} \\ x &< -1 \end{aligned}$$

❸ Samuel and Frank are starting a business making wooden bowls. At the beginning of this year Samuel had already made twenty five bowls. He continues to make eight bowls every month. Frank just started making bowls recently, so he had only made eleven at the beginning of the year. Frank is able to work more quickly, so he makes ten bowls every month. Create a mathematical expression to demonstrate this relationship. During what months this year will Frank have made more bowls than Samuel?

- **Mathematical Relationship:** $25 + 8x < 11 + 10x$ where x is the month of the year.
- **Solve** the inequality to determine the appropriate months. January is considered month 1.

$$\begin{aligned} 25 + 8x &< 11 + 10x \\ -11 - 8x \quad &\;\; -11 - 8x \\ \hline 14 &< 2x \\ x &> 7 \end{aligned}$$

Answer: The months after July, which is the 7th month, will be the months when Frank will have made more bowls than Samuel. The months are August, September, October, November, and December.

Reasoning With Equations & Inequalities

❹ Will has \$70 to spend on CD's. Each CD costs \$12 including sales tax. How many CD's can Will buy?

$12x \leq 70$ Use an inequality for this problem -- He has only \$70 and can only spend that much or less.

$\frac{12x}{12} \leq \frac{70}{12}$ Let x = the number of CD's Will can buy

$x \leq 5\frac{5}{6}$ Notice that the domain is whole numbers since he can't buy a part of a CD.

Answer: Will can buy 5 CD's.

GRAPHING SOLUTIONS

Graphing Simple Inequalities on a Number Line: These are simple inequalities involving one variable, not equations needing lines and shading.

Rules: > or < : Open circle. Line with arrow goes in the appropriate direction.
≥ or ≤ : When = is included, use a closed circle.

The solution set of an inequality with ***one*** < or > is a circle (closed or open as indicated by the problem) with an arrow showing the solution set.

Example $x > -7$ -7 -6 -5 -4 -3 -2 -1 0 1 2 3 $SS = \{x \mid x > -7\}$

The solution set of an inequality with ***two signs*** is where the graphs of the two separate parts overlap. Take the problem apart and graph each part. The solution set is marked where both lines overlap when drawn on a number line.

Example $2 \leq x + 3 < 7$

Solve and graph each part: $2 \leq x + 3$
$-3 \quad -3$

Graph this part of the solution: $-1 \leq x$ *or* $x \geq -1$

Now solve second part: $x + 3 < 7$
$-3 \quad -3$

Graph this part of the solution: $x < 4$

−3 −2 −1 0 1 2 3 4 5 6 7 8

$SS = \{x \mid -1 \leq x < 4\}$ The solution set is determined by analyzing the number line where the 2 graphs have points in common. (In certain cases, there are no points in common and the solution set must be written to show this.) To determine if a given value is a solution to an inequality, simply substitute the value given in the original inequality and see if it checks.

Check: Is 5 a solution to the inequality shown above?

$2 \leq x + 3 < 7$

$2 \leq 5 + 3 < 7$

$2 \leq 8 \leq 7$ False. Since the final result is NOT true, 5 is not a solution to the original inequality.

3.10

<u>Graphing an Inequality on a Coordinate Graph</u>: Inequalities are graphed using the same procedures that are used for graphing equations (See page 74). The line that is graphed becomes the boundary line of the inequality with the points on one side or the other making up the solution set. The border line itself is part of the solution set if ≤ or ≥ is involved.

Steps

1) Boundary lines for inequalities, broken or solid (see #3), are plotted on a coordinate graph like equation lines are graphed. Solve each equation for y using the $y = mx + b$ form. m represents the slope, and b identifies the y-intercept. (Reminder: Reverse the < or > sign if you multiply or divide by a negative number.)
2) Plot the y-intercept point. Locate the next two points using the slope.
3) BROKEN OR SOLID? Draw the line carefully. If an inequality contains < or > (without =) the boundary line is broken. If it contains ≥ or ≤, the boundary line is solid.
4) TEST POINT AND SHADING: Choose any point on the graph that is NOT on the line drawn. This is a test point. Put the values for (x, y) of that point in the original problem. If they make the inequality true, then shade the same side of the graphed line where the test point is located. If the x and y values of the test point make the inequality false, shade the side of the line opposite the test point.
5) Label the graph: Put the original problem on the side of the graph which is shaded - place it near the graph line.

Example On a coordinate plane, graph the following inequality and label its solution set S: $2x + y < 4$

$2x + y < 4$ (*Boundary line will be broken*)

$\underline{-2x \qquad -2x}$

$y < -2x + 4$

slope = –2

y-intercept = 4

Choose a Test Point: (3, 3)

Test it in the original:

$-2x + y < 4$

$2(3) + 3 < 4$?

$9 < 4$ False

Because the test was false, shade the side of the boundary line that is opposite the side where (3, 3) is located. (3, 3) is NOT part of the solution set. Label the shaded part "S" and write the original inequality in that part.

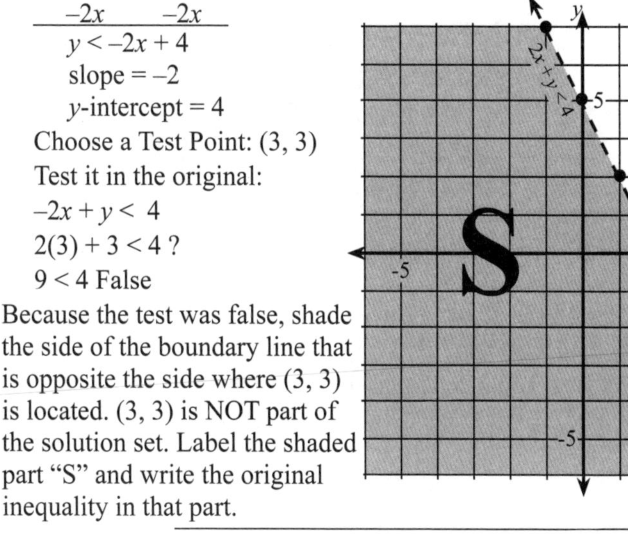

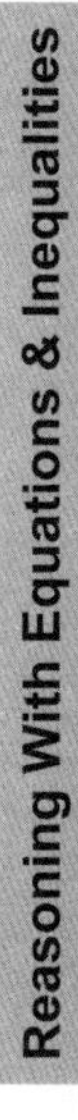

3.11 Systems of Inequalities

Graphing Systems of Inequalities:

1) Graph the first problem using the steps shown on the previous page. Find the test point, shade, and label.
2) Graph the second problem using the same method. This time make the shading go in a different direction.
3) The section of the graph where the shading overlaps is the solution set of the inequality system. All the points in that section are in the solution set and will make both problems true. Mark a large **S** in the section that is the solution set if instructed to do so.
4) If asked to name a point in the solution set, choose any point in that overlapping area and write its coordinates in the (x, y) form. If asked to name a point not in EITHER solution set, choose a point in the section that has NO shading at all and write its coordinates in the (x, y) form.

Examples

❶ Solve graphically. Label the solution set of this pair of inequalities, $y \geq x$ and $y < -x - 2$, with **S**. Give the coordinates of a point in the solution set.

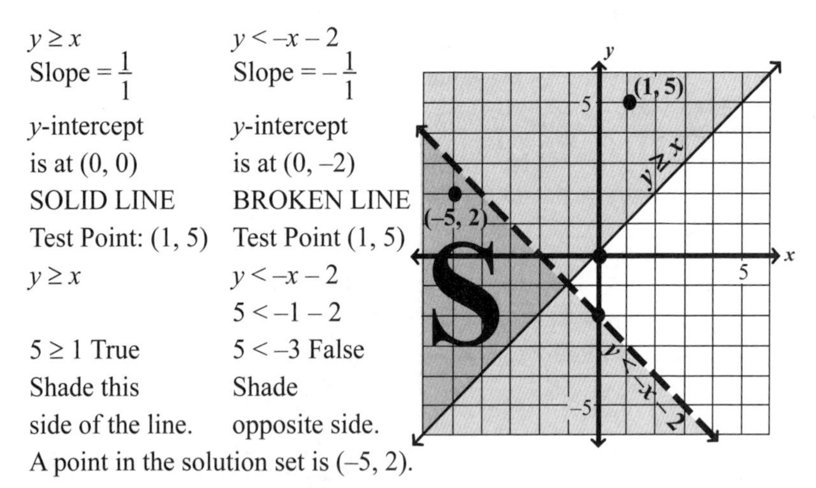

$y \geq x$	$y < -x - 2$
Slope $= \frac{1}{1}$	Slope $= -\frac{1}{1}$
y-intercept is at (0, 0)	y-intercept is at (0, –2)
SOLID LINE	BROKEN LINE
Test Point: (1, 5)	Test Point (1, 5)
$y \geq x$	$y < -x - 2$
	$5 < -1 - 2$
$5 \geq 1$ True	$5 < -3$ False
Shade this side of the line.	Shade opposite side.

A point in the solution set is (–5, 2).

Notes: • The solid line needed to graph $y \geq x$ indicates that the points on the border line are included in the solution set.

• INEQUALITY coordinate graphs do require test points and shading. Do not confuse inequalities with equations.

❷ Juan is serving on the prom committee and has been put in charge of snacks. He wants to buy an assortment of cheeses and crackers. One package of gourmet crackers costs \$3.00. The cheese costs \$5.00 per pound. He has only been allotted \$50.00 total for the snacks so he wants to spend less than that amount. From talking to his mom, Juan knows that he will buy at least 4 pounds of cheese to serve his classmates. Create the mathematical representation of this problem and find two possible combinations of boxes of crackers and pounds of cheese that he can buy.

Solution: This problem needs to be solved using inequalities. One inequality represents the total possible cost less than \$50 and the other represents the purchase of at least 4 pounds of cheese. Graph both in equalities to find the solution set and choose two points within the solution set as the answers.

Let x = boxes of crackers
y = pounds of cheese
$3x + 5y < 50$
$y \geq 4$

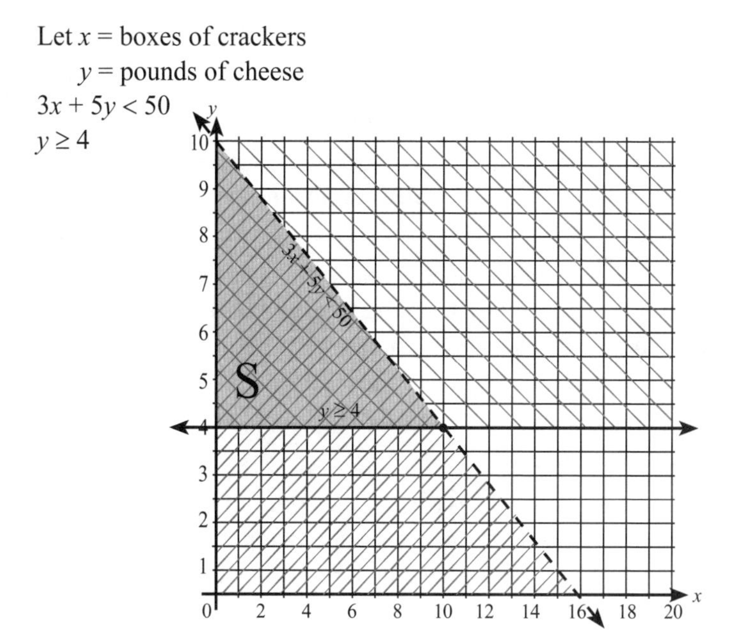

Conclusion: (6, 4) and (4, 5) are both within the solution sets. He could buy 6 boxes of crackers and 4 pounds of cheese, or he could buy 4 boxes of crackers and 5 pounds of cheese. (Note: There are multiple combinations that could result in having 4 or more pounds of cheese and spending less than \$50. The coordinates (x, y) of any point within the solution set could be substituted for crackers and cheese respectively in the inequalities with acceptable results.)

Reasoning With Equations & Inequalities

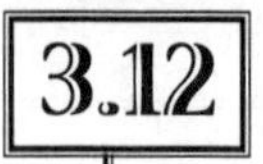

SOLVING LINEAR PROBLEMS WITH CONSTRAINTS

Constraint: A restriction, boundary, or limit that regulates possible outcomes.

Working with linear polynomials can help find the possible maximum or minimum values in problem solving. Systems of linear inequalities are used in real world situations where it is necessary to combine several resources to produce a maximum or minimum result. Using linear inequalities provides a feasible region which shows all the possible outcomes of the process based on an objective equation or function which defines the purpose of the problem. These outcomes are restricted by the boundary lines of the inequalities. The restrictions are called constraints. The maximum and minimum output values are found at the intersections of the lines that are the boundaries of the inequalities. This process is also called linear programming.

Steps

1) Determine what the purpose of the problem is and write a linear equation to represent it. This equation is often called the objective equation or the optimization equation.
2) Write inequalities that represent the desired solution of each component of the problem. These are called restrictions, or constraints.
3) Graph each inequality on a coordinate plane and indicate the correct solution areas. Each inequality line is the boundary of that part of the solution. The portion of the graph where all of the solutions overlap is called the feasible area or region.
4) Use the linear equations as a system of equations to find the points of intersection of the boundary lines. These points will be the corners of the feasible region.
5) Substitute the coordinates each of the points of intersection in the "Objective Equation." The coordinates that provide the highest or lowest value when substituted in the objective equation are the x and y values of the solution.
6) Write a conclusion.

The first example on the next page demonstrates the development of a problem where the individual parts of the problem are already defined as mathematical statements. In the second example, the necessary mathematical statements must be developed from an application.

Example

❶ Maximize and minimize this equation given the restrictions listed.

Steps *Note:* Steps **1)** and **2)** are not needed in this example. Start with step 3.

Objective Equation: $F(x, y) = x + 1.5y$

Constraints:
$$y - x \geq 2$$
$$x + y \leq 15$$
$$2y + 4x \leq 36$$
$$x \geq 0$$
$$y \geq 0$$

3) Rewrite each inequality to slope intercept form for easier graphing. Graph on one coordinate plane. Locate the feasible area where there is a common solution (all the solutions overlap).

$$y \geq x + 2$$
$$x + y \leq 15; \quad y \leq -x + 15$$
$$y \leq \frac{(-4x + 36)}{2}; \quad y \leq -2x + 18$$
$$x \geq 0; \quad y \geq 0$$

Note: Only the feasible area is shaded on this graph.

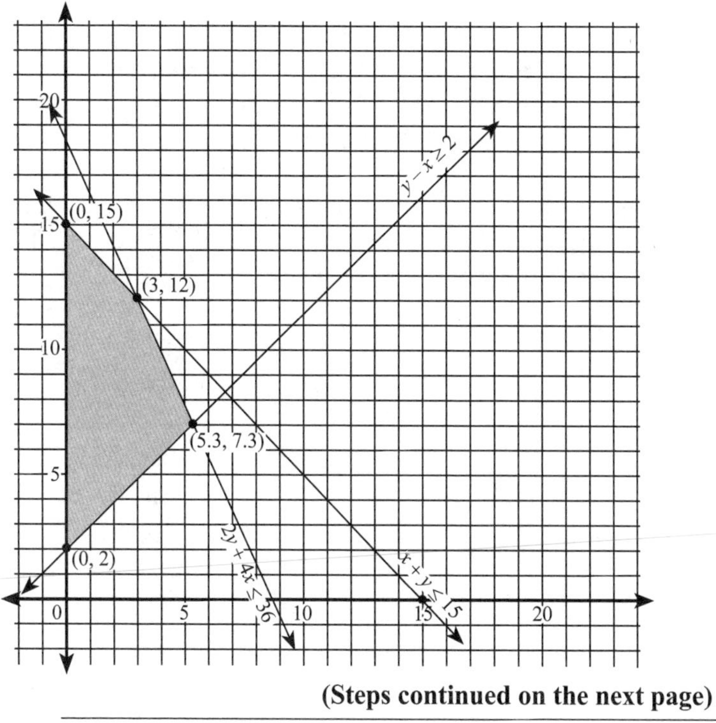

(Steps continued on the next page)

Reasoning With Equations & Inequalities

4) Find the coordinates of the points of intersection of each of the 4 corners of the feasible region.

1) $x = 0$ and $y = -x + 15$; (0, 15)

2) $x = 0$ and $y = x + 2$; (0, 2)

3) $y = -2x + 18$
$\underline{-(y = x + 2)}$
$0 = -3x + 16$

$x = \frac{16}{3}$ or 5.3

$y = x + 2$; $y = 7.3$
Intersection: (5.3, 7.3)

4) $y = -2x + 18$
$\underline{-(y = -x + 15)}$
$0 = -x + 3$
$x = 3$
$y = -(3) + 15 = 12$
Intersection: (3, 12)

5) Test the 4 pair of coordinates in the objective equation.
$F(x, y) = x + 1.5y$

$F(0, 15) = 0 + 1.5(15) = 22.5$ Maximum output
$F(0, 2) = 0 + 1.5(2) = 3$ Maximum input
$F(5.3, 7.3) = 5.3 + 1.5(7.3) = 16.25$
$F(3, 12) = 3 + 1.5(12) = 21$

6) Conclusion: The maximum output for this problem is 22.5 which is reached at the intersection of $x \geq 0$ and $y \leq -x + 15$. The minimum output is 3 which is at the intersection of $x \geq 0$ and $y \geq x + 2$.

Example

❷ Jessie is working at her bakery making treats for the local market. Each batch of 12 brownies costs her \$6.00 to produce and each batch of 12 cookies costs about \$8.00 to produce. She has a budget of \$100 available to purchase the ingredients. One batch of cookies takes 30 minutes to make, and one batch of brownies takes 45 minutes to make. She has ten hours to do the baking. In order to optimize the profit, how many batches of each should she make if she charges \$10.00 for a box of 12 cookies and \$20.00 for a box of 12 brownies.

Analysis: The constraints or restrictions here are the ten hours for baking and the \$100 allowed for ingredients. Set up the "let statements" and determine how much profit will be made on each batch of brownies and cookies. Notice that the baked goods will be sold in boxes that each contain one batch of either brownies or cookies. Then write the equation for the final profit to be optimized so Jessie can make as much profit as possible.

Steps

1) Let x = batches of brownies
y = batches of cookies

Profit: 1 batch of Brownies (x): \$20 – \$6 = \$14
1 batch of Cookies (y): \$10 – \$8 = \$2

Optimize Final Profit: $P = 14x + 2y$

2) Now write the inequalities the represent the constraints or restrictions on the problem.

$6x + 8y \le 100$	Cost of x batches of brownies and y batches of cookies.
$0.75x + 0.5y \le 10$	Hours for baking:
$x \ge 0$; $y \ge 0$	Quantities of brownies and cookies cannot be negative.

3) Solve and graph them all on one coordinate grid.

Time	Cost	Quantities
$0.75x + 0.5y \le 10$	$6x + 8y \le 100$	$x \ge 0$
$y \le \dfrac{-0.75x + 10}{0.5}$	$y \le \dfrac{-6x + 100}{8}$	$y \ge 0$
$y \le -1.5x + 20$	$y \le -0.75x + 12.5$	

(Steps continued on the next page)

For clarity, only the overlapping area, called the feasible area, is shaded on this graph.

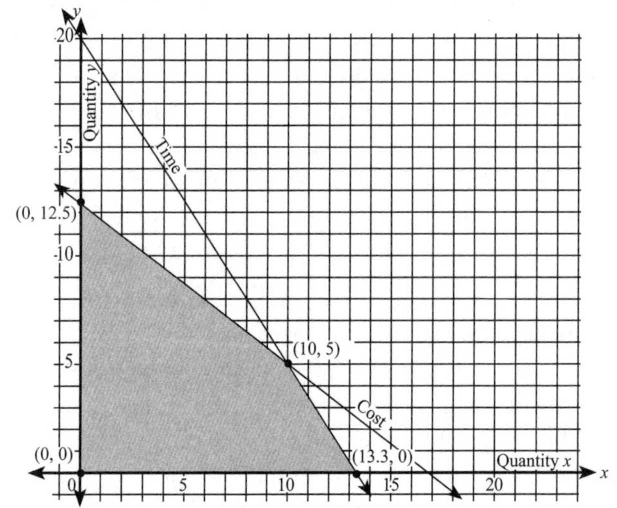

4) Interpret the boundary line created by the inequalities of the feasible area as equations for this step. Find the points of intersection of each pair of boundary lines. These are the "corners" of the feasibility area. The maximum profit will be attained at one of them.

Time & Cost

$y = -1.5x + 20$

$-(y = -0.75x + 12.5)$

$0 = -0.75x + 7.5$

$0.75x + 7.5$

$x = 10$

$y = -1.5(10) + 2$

$y = 5$

Time & Cost: (10, 5)

Time & x Quantity

(x intercept)

$0.75x + 0.5y = 10$

$0.75x + 0.5(0) = 10$

$x = \dfrac{10}{.75} = 13.3$

(13.3, 0)

Cost & y Quantity

(0, 12.5) (y-intecept)

Quantities x and y: (0, 0)

There are 4 corners of the feasible area at the intersections of the constraints. They are: (0, 0), (13.3, 0) (10, 5), (0, 12.5)

5) The x and y coordinates of each point must be tested in the Final Profit equation to find the optimal profit.

6) Conclusion: For the maximum profit, Jessie should bake 13 batches of brownies and no cookies.

- Another Question: How could Jessie maximize her profit if she must include both brownies and cookies in the sale?
- Answer: If she bakes 10 batches of brownies and 5 batches of cookies her profit will be $150.

3.13

Quadratic Equations

See also Quadratic Functions – Unit 4

Quadratic: An equation which contains a "squared" variable, like x^2 is called a quadratic equation. Quadratic equations can be solved algebraically or graphically. The standard form of a quadratic equation is $ax^2 + bx + c = y$. The graph of the quadratic equation is a parabola.

Roots or Zeros: Quadratic equations may have one, two, or no real roots or zeros. The roots of the quadratic equation are the points where $y = 0$ which is where the graph of the equation crosses the x-axis. If only one root exists, the graph of the parabola is tangent to the x-axis. If no real roots exist the equation does not intersect with the x-axis at all. Sometimes the roots are equal. A root can be 0. Both roots will check in the original equation if they are correct. Some roots are mathematically correct but do not fit the requirements of the problem. They must be marked as rejects.

Solutions: The values of x and the corresponding value of y which make the equation true.

Finding The Roots Algebraically: Factor using square roots, the quadratic formula, or completing the square.

Factoring: Steps

1) Move all the terms to one side of the equal sign using algebraic methods and if needed, replace y with 0.
2) Write the terms (usually located on the left side of the equal sign for ease in working) in standard form.
3) The other side of the equal sign is 0. (Usually the right side.)
4) If a variable with an exponent of "2" is still in the problem, you must factor. Remember to look for a common factor first, then factor the remaining equation using ( ) ( ). See Unit 3.4 for review of factoring.
5) KEEP the = 0. This is an equation and it must have an equal sign.
6) Separate the factors and make each set of ( ) = 0. This utilizes the zero product property - also called the multiplication property of zero. This property says that if a product equals zero, one or more of its factors must be zero.
7) Solve each equation for the variable.
   Note: The roots are the opposite numbers from those in the factors.
8) Check each answer in the original quadratic equation. Both answers must check.
9) Indicate whether both answers are to be used or if one is a reject based on the information in the problem. A reject must be clearly marked "reject".

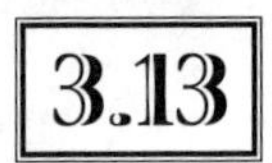

Examples **Solve By Factoring**

❶ Equation: $x^2 - 14 = -5x$

Steps:

1) Move all terms to one side of equal sign: $x^2 + 5x - 14 = 0$

2) Factor the equation: $(x + 7)(x - 2) = 0$

3) Separate the factors: $(x + 7) = 0 \mid (x - 2) = 0$

4) Solve both equations for x: $x = -7 \mid x = 2$

5) Check both answers in $x^2 + 5x - 14 = 0$:

$$(-7)^2 + 5(-7) - 14 = 0$$
$$49 - 35 - 14 = 0$$
$$0 = 0\checkmark$$

6) Both check:

$$(2)^2 + 5(2) - 14 = 0$$
$$4 + 10 - 14 = 0$$
$$0 = 0\checkmark$$

Note: The roots are the opposite numbers that are in the factors. If the equation above is graphed, the graph will intersect the x-axis at -7 and at 2.

❷ $3x^2 + 12x = 15$

$3x^2 + 12x - 15 = 0$

$3(x^2 + 4x - 5) = 0$

$3(x + 5)(x - 1) = 0$

$x + 5 = 0$; $x - 1 = 0$

$x = -5$ $\quad x = 1$

$x = -5$ and $x = 1$

(The check is left to the student.)

Square Roots: When solving an equation like $x^2 - 36 = 0$, (there is no "x" term in the middle) just add "36" to both sides. Then take the square root of each side.

Examples

❶ $x^2 - 36 = 0$

$\quad +36 \quad +36$

$x^2 = 36$

$x = 6$ and -6 (Remember to check both answers in the original.)

❷ $x^2 - 98 = 0$

$x^2 = 98$

$x = \pm\sqrt{98}$

$x = \pm 7\sqrt{2}$

$x = 7\sqrt{2}$ and $x = -7\sqrt{2}$

3.13

Quadratic Formula: This formula is used commonly at higher levels of math to solve quadratic equations. The equation is put in standard form and the coefficients are assigned letters that are used in the formula. Identify the value of a, b, and c before starting.

$$x = \frac{-b \pm \sqrt{b^2 - 4ac}}{2a}$$

Deriving the Quadratic Formula

This formula is widely used to solve quadratic equations, and parts of it are used for other mathematical calculations. In addition to knowing how to use the formula, it is important for the students to be familiar with its derivation.

Steps

1) Put the equation in standard form. $ax^2 + bx + c = 0$
2) Move the constant to the other side of the equation. $\quad -c \quad -c$; $ax^2 + bx = -c$
3) Divide each term by the coefficient of the term that is squared. $x^2 + \frac{b}{a}x = \frac{-c}{a}$
4) Divide the coefficient of the x term by 2 and square it. When the quotient is squared, it will be positive at this level of mathematics.* $\left[\frac{b}{a} \div 2 = \frac{b}{2a}\right]$ $\left(\frac{b}{2a}\right)^2 = \frac{b^2}{4a^2}$
5) Add the squared term to both sides of the equation. $x^2 + \frac{b}{a}x + \frac{b^2}{4a^2} = \frac{-c}{a} + \frac{b^2}{4a^2}$
6) Simplify the right side of the equation using a common denominator. Write the result as a single fraction. $x^2 + \frac{b}{a}x + \frac{b^2}{4a^2} = \frac{-c(4a)}{4a^2} + \frac{b^2}{4a^2}$ $x^2 + \frac{b}{a}x + \frac{b^2}{4a^2} = \frac{-4ac + b^2}{4a^2}$
7) Factor the left side of the equation into two equal factors. Write them as the square of a binomial. $\left(x + \frac{b}{2a}\right)^2 = \frac{b^2 - 4ac}{4a^2}$
8) Take the square root of both sides of the equation. $x + \frac{b}{2a} = \pm\sqrt{\frac{b^2 - 4ac}{4a^2}}$
9) Solve for x. $x + \frac{-b}{2a} \pm \frac{\sqrt{b^2 - 4ac}}{2a}$
10) Simplify unless indicated otherwise. ** $x = \frac{-b \pm \sqrt{b^2 - 4ac}}{2a}$

Notes: * When working with complex numbers in higher math courses, step 4 may result in a non-positive number.

** The right side of the equation is not always combined into a single fractions as it is in step 10 above.

Reasoning With Equations & Inequalities

❶ Solve: $x^2 + 4x = 14$

Write in Standard Form: $x^2 + 4x - 14 = 0$

Identify a, b, and c: $a = 1,\ b = 4,\ c = -14$

Write the Quadratic Formula: $x = \dfrac{-b \pm \sqrt{b^2 - 4ac}}{2a}$

Substitute: $x = \dfrac{-(4) \pm \sqrt{(4)^2 - 4(1)(-14)}}{2(1)}$

Do the calculations: $x = \dfrac{-4 \pm \sqrt{72}}{2}$

Simplify the radical: $x = \dfrac{-4 \pm 6\sqrt{2}}{2}$

Simplify the fraction: $x = -2 \pm 3\sqrt{2}$

Roots or Zeros: $x = -2 + 3\sqrt{2}$ *and* $x = -2 - 3\sqrt{2}$

❷ $2x^2 - 7x + 4 = 0$

$a = 2,\ b = -7,\ c = 4$

$$x = \frac{-b \pm \sqrt{b^2 - 4ac}}{2a}$$

$$x = \frac{-(-7) \pm \sqrt{(-7)^2 - 4(2)(4)}}{2(2)}$$

$$x = \frac{7 \pm \sqrt{49 - 32}}{4}$$

$$x = \frac{7 \pm \sqrt{17}}{4}$$

$$x = \frac{7 + \sqrt{17}}{4},\ x = \frac{7 - \sqrt{17}}{4}$$

The answers as given are exact answers. If a problem asks for answers to the nearest hundredth, for example, or if you want to graph the equation using an approximation of the roots, it is necessary to calculate the decimal value of the roots using the calculator.

$x = \dfrac{7 + \sqrt{17}}{4} \approx 2.78$ and $x = \dfrac{7 - \sqrt{17}}{4} \approx 0.72$

Note: In examples 1 and 2, the check is left to the student.

Completing The Square is another method of solving quadratic equations. It makes an unfactorable equation into an equation which can then be solved by factoring.

- If $a = 1$, solve by **completing the square**. If $a \neq 1$, divide each term in the equation by a before beginning. It is often more efficient to use the quadratic formula when $a \neq 1$. (See also page 97)

Examples

❶ $x^2 - 6x + 2 = 0$

Steps

1) Subtract c from both sides: $x^2 - 6x = -2$

2) Divide b by 2 and square it: $b = -6 : \left(\frac{-6}{2}\right)^2 = (-3)^2 = 9$

3) Add to BOTH sides of the equation: $x^2 - 6x + 9 = -2 + 9$

4) Factor the left side into 2 equal factors: $(x - 3)(x - 3) = 7$

5) Take the square roots of both sides: $(x - 3) = \pm\sqrt{7}$

6) Solve for x: $x = 3 + \sqrt{7}, \quad x = 3 - \sqrt{7}$

Solution $= \{3 + \sqrt{7}, 3 - \sqrt{7}\}$

❷ $x^2 - 3x + 1 = 0$

$x^2 - 3x = -1$

$b = (-3) : \left(\frac{-3}{2}\right)^2 = \left(\frac{9}{4}\right)$

$x^2 - 3x + \frac{9}{4} = -1 + \frac{9}{4}$

$\left(x - \frac{3}{2}\right)^2 = \frac{5}{4}$

$x - \frac{3}{2} = \pm\sqrt{\frac{5}{4}}$

$x = \frac{3}{2} + \frac{\sqrt{5}}{2} = \frac{3 + \sqrt{5}}{2}; \quad x = \frac{3}{2} - \frac{\sqrt{5}}{2} = \frac{3 - \sqrt{5}}{2}$

Solution $x = \frac{3 + \sqrt{5}}{2}$ and $x = \frac{3 - \sqrt{5}}{2}$

Note: The check is left to the student.

Rejects: Negative answers for distance or age; fractions and decimals if the domain is integers; positive answers if the domain is negatives; odd numbers if the domain is evens, etc. REREAD the problem carefully before deciding if there are any rejects.

Example

The length of a rectangle is 2 more than its width. The area of the rectangle is 35. Find its dimensions.

Steps:

1) Make a diagram: (rectangle with sides x and $x + 2$)
2) Formula for Area = $l \cdot w$
3) Equation: $(x)(x + 2) = 35$
4) Simplify: $x^2 + 2x = 35$

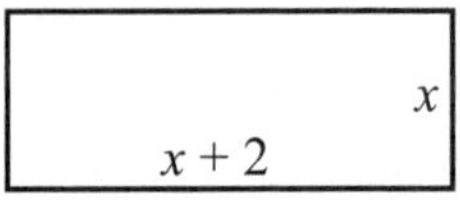

$$\underline{\quad -35 \quad -35}$$

$$x^2 + 2x - 35 = 0$$

5) Factor: $(x - 5)(x + 7) = 0$
6) Solve: $x - 5 = 0$ | $x + 7 = 0$

$x = 5$ | $x = -7$ (Reject -7 because a rectangle cannot have a negative side.)

7) Find the other side: $x + 2 = 7$
8) Answer: The rectangle is 5 by 7.
9) Check: $(5)(7) = 35$

Functions

- Understand the concept of a function and use function notation.
- Interpret functions that arise in applications in terms of the context.
- Analyze functions using different representations.
- Build a function that models a relationship between two quantities.
- Build new functions from existing functions
- Construct and compare linear, quadratic, and exponential models and solve problems.

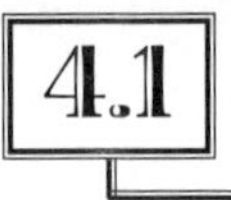

RELATIONS AND FUNCTIONS

Relations and **functions** are terms that can be used to name equations or rules in higher math. Both refer to a set of ordered pairs that are associated to each other in a way that is expressed by the relation or the function. Functions and relations both have domains and ranges.

Element: A member of the domain or range.

Relation: A set of ordered pairs that have a connection to each other that is defined by the relation or rule. The ordered pairs may be numeric or not.

Examples ❶ (2, 4), (6, 8), (6, 4)

❷ (Albany, NY), (Boston, MA), (Sacramento, CA)

Function: A function is a relation that consists of a set of ordered pairs in which each value of x is connected to a unique value of y based on the rule of the function. For each x value, there is one and only one corresponding value of y.

Example (2, 3), (3, 4), (4, 5), (5, 4), (6, 4)

Note: No x's are repeated, but y can be repeated.

Function Notation is $f(x)$. Any letter can name the function, and the independent variable is noted in the ( ). For instance, $h(x)$, or $g(x)$. This notation replaces "y" in the usual equation format. Instead of $y = x + 3$, it is $f(x) = x + 3$. The output of the function for the element of the domain is $f(x)$.

Note: Remember that functions have certain characteristics and the function notation cannot be applied to equations that do not represent functions. So if you see $f(x)$ you know that the function definition will be true for that equation. In a function, the set of values for the independent variable, x, is called the domain. It is mapped (connected) to the set of values for the dependent variable, $f(x)$ or y, called the range. The function itself is the rule that makes that connection. There are many ways to write this.

The function, $f(x) = x^2 + 4$ can also be written in the following ways:

1) $f(x) = x^2 + 4$ The output value of the function paired with x is $x^2 + 4$.
2) $y = x^2 + 4$ The output value, y, is paired with x in this function.
3) $f = \{(x, y) \mid y = x^2 + 4\}$ or $f = \{(x, y) : y = x^2 + 4\}$ This indicates that the function is the set of values for x and y that fulfill the requirements of the rule. Either | or : is used to mean "such that".

4.1

Domain: The largest set of elements available for input as the independent variable, the first member of the ordered pair (x). The domain is the set of Real Numbers unless otherwise noted. Restrictions on the domain can be found in problems involving fractions, radicals, word problems, and when a domain is specified.

- **Fraction:** Denominator cannot be zero. $f(x) = \frac{x-4}{x+3}, x \neq -3$
- **Radical:** Radicand cannot be negative. $f(x) = \sqrt{4x-10}; x \geq \frac{5}{2}$
- **Word Problems:** People ≥ 0; Distance ≥ 0
- **Problem Specifications:** $5 \leq x \leq 20$
- **Reasonable Domain:** A distance cannot be negative, so the domain for distance problems is the non-negative numbers.

Range: The set of elements available for the dependent variable or output; the second member of the ordered pair (y). The range can be found for the entire relation or function, or it can be found for specific domains. If a value of x is given as the domain, substitute to find the corresponding value of y (range).

Working With the Domain and Range

Examples

❶ Find the domain and range of the equation $f(x) = x + 4$. The domain (values available for x) is R, (Real Numbers). The range in this case (numbers available for y) is also R. (There are no restrictions on this function.)

❷ $y = x^2$ Domain is the Reals, range is $y \geq 0$.
(Any real number squared is positive or zero.)

❸ If the domain is $\{-3, 0, 4\}$ find the range of $y = x - 3$

Substitute -3: $y = (-3) - 3 = -6$

Substitute 0: $y = (0) - 3 = -3$

Substitute 4: $y = 4 - 3 = 1$

Note: In this relation, if the domain is $\{-3, 0, 4\}$ then the range is $\{-6, -3, 1\}$

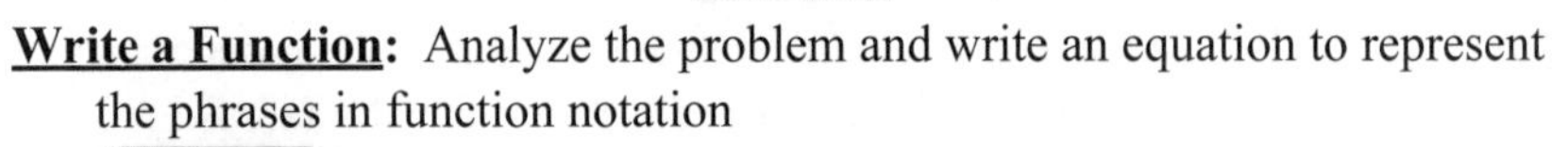

4.1

Write a Function: Analyze the problem and write an equation to represent the phrases in function notation

Example Write a function for the following situation: The output value of the function is three more than four times the input value.
$f(x) = 4x + 3$

Example At a charity event, a donation of \$500.00 is received. Additional income is based on the number of tickets sold. Write a function to represent the total amount raised if each ticket, t, is sold for \$32.00.
$f(t) = 32t + 500$

Evaluate a Function: Specific values of x are substituted in the function to find the output value. Write the x value in the function name, then substitute and do the math.

Using the example above, find the value of $f(x)$when $x = 6$: $f(6) = 4(6) + 3 = 27$

Now find the value of $f(x)$ when x is -5: $f(-5) = 4(-5) + 3 = -17$

Example Find the values of $f(x)$, $g(x)$, and $h(x)$ when $x = 4$, 5, and 6. Write the solutions as ordered pairs.

$f(x) = x - 5$	$g(x) = 2x + 5$	$h(x) = x^2$
$f(4) = 4 - 5 = -1$	$g(4) = 2(4) + 5 = 13$	$h(4) = (4)^2 = 16$
$f(5) = 5 - 5 = 0$	$g(5) = 2(5) + 5 = 15$	$h(5) = (5)^2 = 25$
$f(6) = 6 - 5 = 1$	$g(6) = 2(6) + 5 = 17$	$h(6) = (6)^2 = 36$
$(4, -1), (5, 0), (6, 1)$	$(4, 13), (5, 15), (6, 17)$	$(4, 16), (5, 25), (6, 36)$

Note: The solutions are not always written as ordered pairs. Make sure to read the directions. The question might ask "What is the image of 4 in $f(x)$"? The answer would be -1.

or in $g(x)$, what value is x mapped to if $x = 6$? The answer would be 17.

Graph of a Function: When graphing, $f(x)$ becomes the y value of the point to be plotted with the corresponding x. Substitute y for $f(x)$ when graphing.

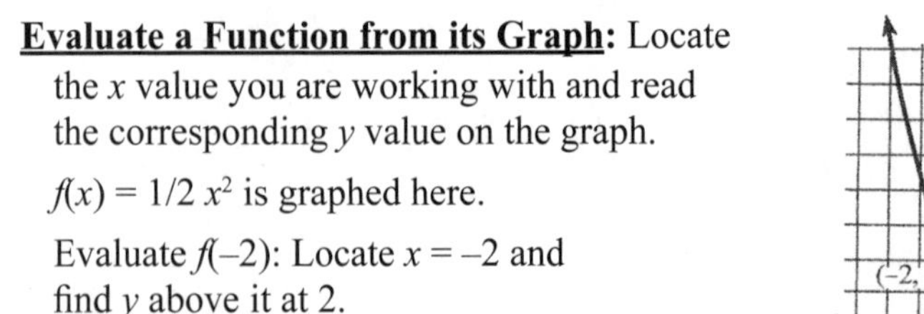

Evaluate a Function from its Graph: Locate the x value you are working with and read the corresponding y value on the graph.

$f(x) = 1/2\ x^2$ is graphed here.

Evaluate $f(-2)$: Locate $x = -2$ and find y above it at 2.

Evaluate $f(4)$: Locate $x = 4$, find y above it at 8.

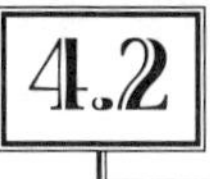

4.2 GRAPHS OF FUNCTIONS

Determining if a Graph is a Function: Since relations and functions are both sets of ordered pairs, we can graph them and compare them on a graph. A test called the "vertical line test" can be used to determine if a graph is a relation or a function. In this test, a vertical line (like the edge of a pencil) is moved across the graph from left to right.

- **Vertical Line Test:** When a function is graphed, a vertical line passed across the graph will intersect the function graph in only one point at a time. A function has only one unique value of y for any value of x in the domain.

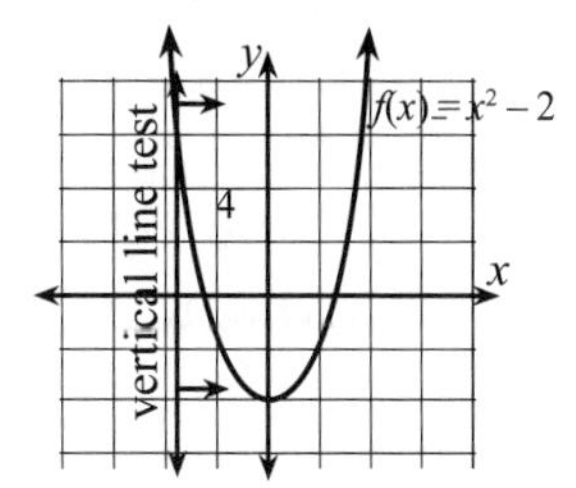

If the graphed line is intersected by the vertical line in more than one place at a time, the graph is a *relation*. If the vertical line intersects the graph in only point at a time, the graph is a function.

Examples **Determine whether the graph is a function or relation.**

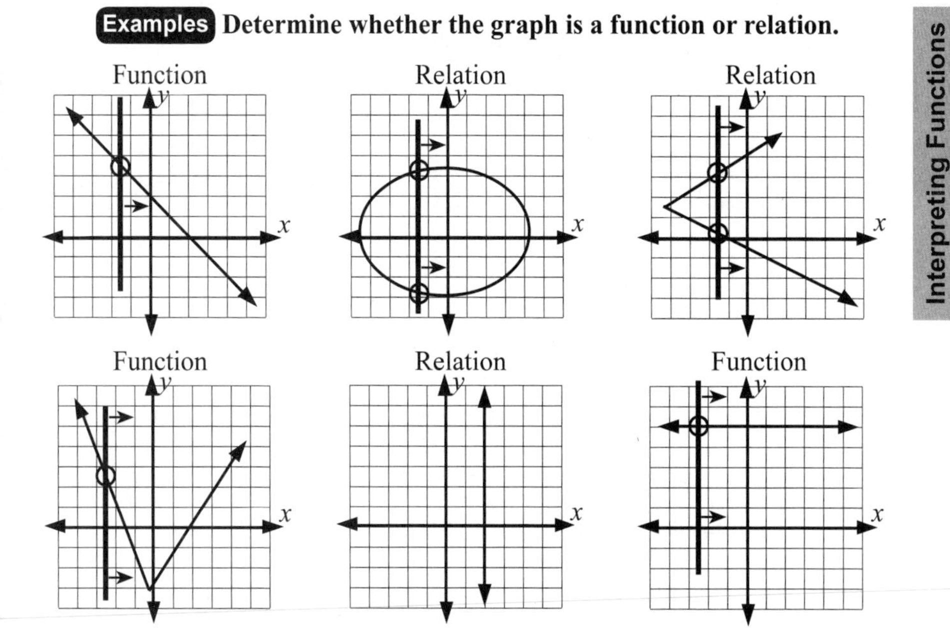

4.2

One to One Function: 1-1 function. First it must be a function. Secondly, when the ordered pairs are examined, no two of them have the same y value. No x's can be repeated, and no y's can be repeated.

In a one-to-one function, a vertical line test works, and also a horizontal line passed over the graph will intersect the graph in only one point at a time.

Examples

❶ (2, 3), (3, 4), (4, 5) is 1 – 1.

❷ (2, 3), (4, 3), (5, 6) is *not* 1 – 1 because 3 is repeated as a y value.

❸ In the figure to the right, the curve represents $f(x)$. It is 1 – 1 because the vertical line test and the horizontal line test both work.

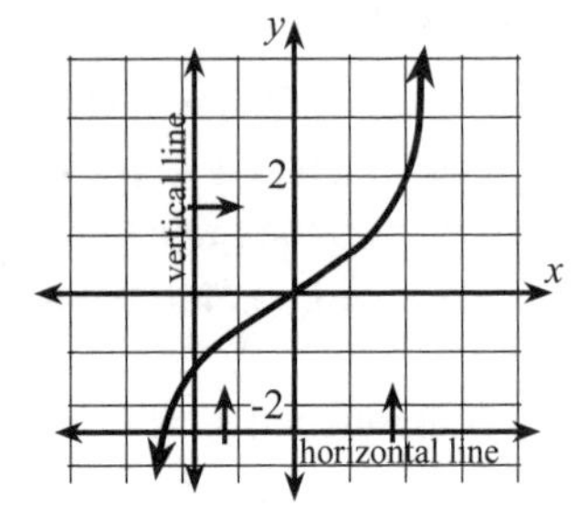

Practice: Find the domain and the range where possible. It is often helpful to use a graph to find the range. Determine if the equation is a function or not, and if it is a function, is it 1-1?

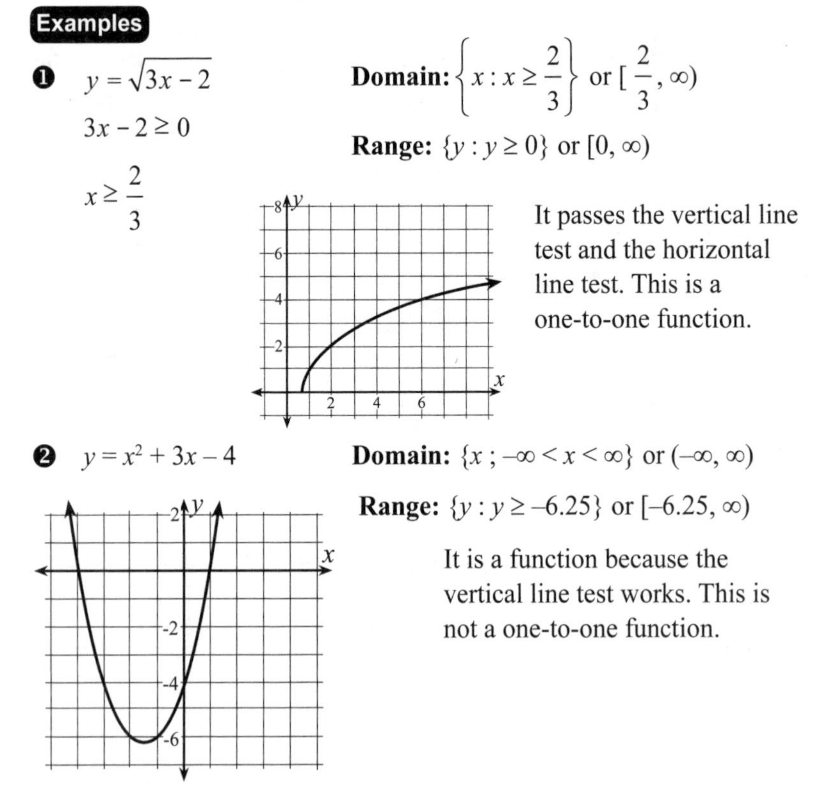

Examples

❶ $y = \sqrt{3x - 2}$

$3x - 2 \geq 0$

$x \geq \dfrac{2}{3}$

Domain: $\left\{x : x \geq \dfrac{2}{3}\right\}$ or $[\dfrac{2}{3}, \infty)$

Range: $\{y : y \geq 0\}$ or $[0, \infty)$

It passes the vertical line test and the horizontal line test. This is a one-to-one function.

❷ $y = x^2 + 3x - 4$

Domain: $\{x ; -\infty < x < \infty\}$ or $(-\infty, \infty)$

Range: $\{y : y \geq -6.25\}$ or $[-6.25, \infty)$

It is a function because the vertical line test works. This is not a one-to-one function.

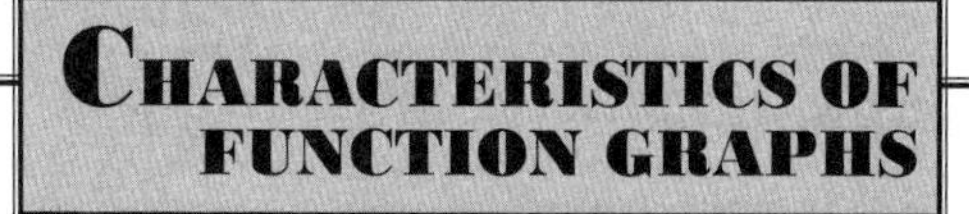

Type of Function: There are many different types of functions. The types of functions discussed here include constant, linear or identity, quadratic, exponential, square root, cube root, absolute value, and piecewise-defined including step functions.

Types of Graphs: There are certain shapes of graphs that are associated with various functions. The general shape of these graphs will be similar although the size and exact shape may vary based on the specific numbers used in the function. The basic functions are called *parent functions*. It is easy and fun to use a graphing calculator to change the size, shape, and location of the graphs of these functions by changing the coefficients of the variables and making other changes in the basic function.

Domain: Remember the domain is the set of values of x that are available for input to the function.

Range: The range is the set of possible output values, y or $f(x)$.

Increasing: For any x_1 and x_2 where $x_1 < x_2$, if $f(x_1) < f(x_2)$ then the graph is increasing in the interval between those points.

Decreasing: For any x_1 and x_2 where $x_1 < x_2$, if $f(x_1) > f(x_2)$ then the graph is decreasing in the interval between those points.

Constant: For any x_1 and x_2 where $x_1 < x_2$, if $f(x_1) = f(x_2)$ then the graph is constant in the interval between those points. This would be a horizontal line.

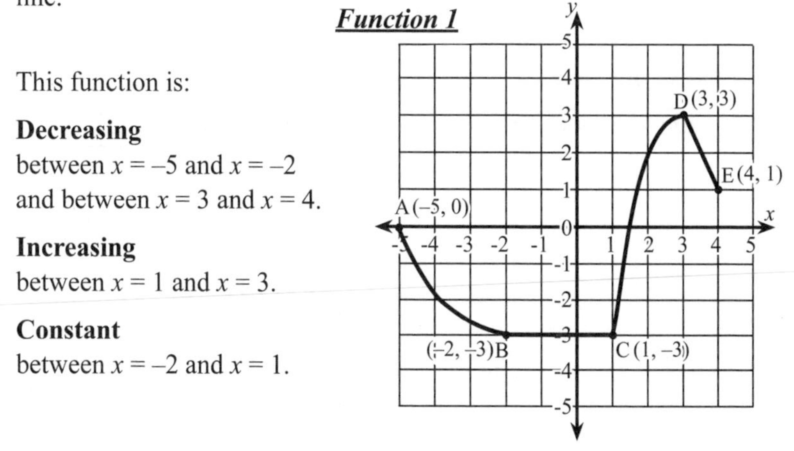

4.3

Vertex: In a parabola, the highest or lowest point. Also called the turning point.

Maxima or Minima: The value of $f(x)$ at the vertex. A maximum point is the point at which the graph changes increasing to decreasing. The mimima, or minimum point, is where the graph changes from decreasing to increasing.

Relative Maxima or Minima: If a graph has more than one turning point, relative maxima or minima are the values of $f(x)$ at the points where the graph changes from decreasing to increasing and vice versa.

Symmetry: Some functions, but not all, will be symmetric to a line.

Intercepts: The point(s) at which the graph intersects the x-axis or y-axis.

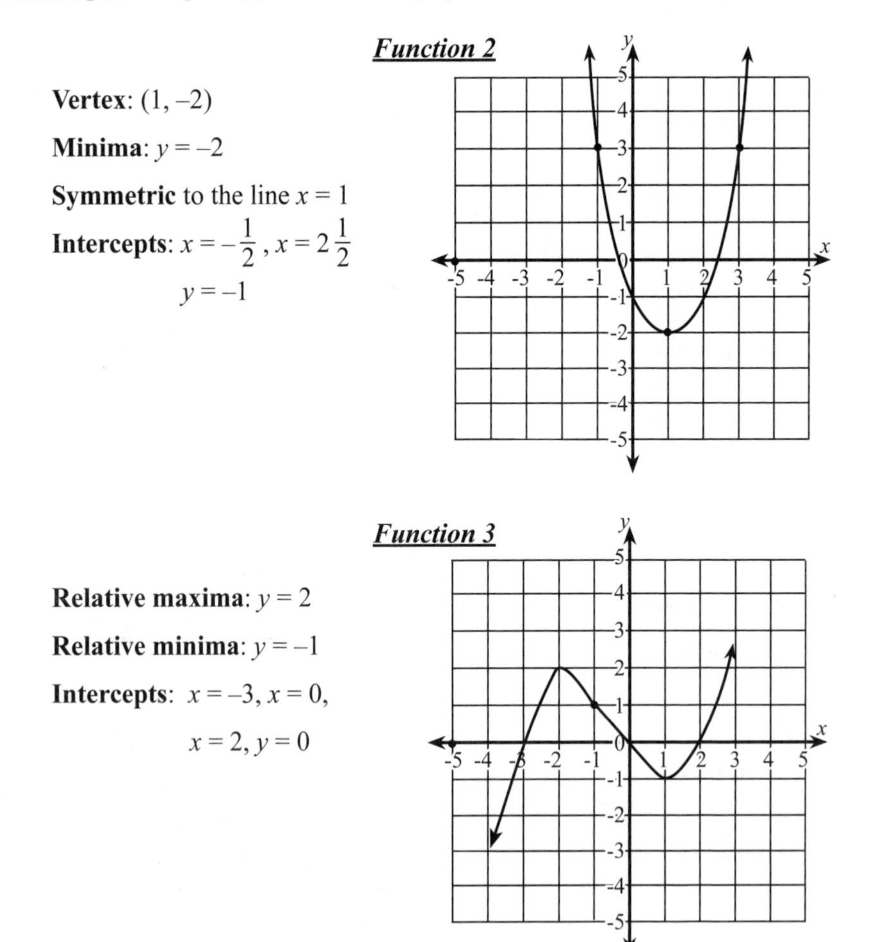

4.3

Asymptote: A line that is approached by a graph as it approaches infinity.

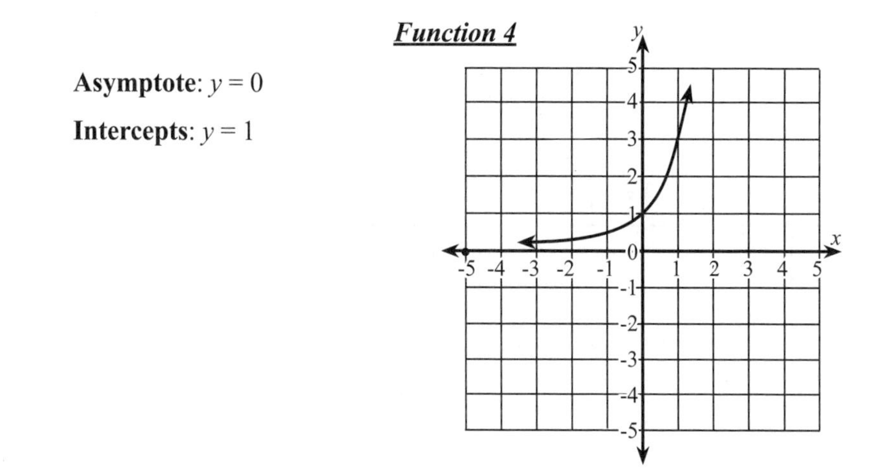

Evaluate: If asked to evaluate $f(x)$ from a graph, locate the given x value on the horizontal axis and read the corresponding y value. In function 4, evaluate $f(1)$. At $x = 1$, $y = 3$. Therefore, $f(1) = 3$.

Parent Function: The basic form of a particular type of function. Examples given are the parent function where possible.

Note: Points read from a graph are approximated. Determine the numbers as closely as possible.

Linear, Quadratic and Exponential Parent Functions

Type of $f(x)$	Parent Function	Sketch	Intercepts	Max/Min	Notes
Constant	$f(x) = b$ ***Note:*** $f(2)$ is used for the sketch.	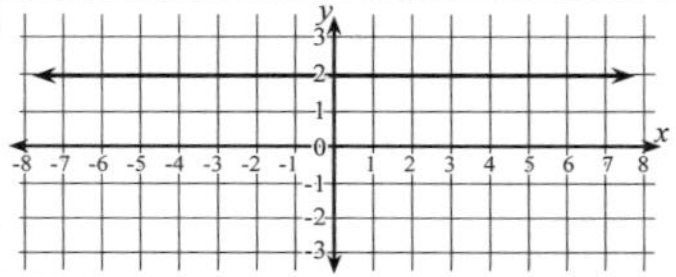	y-intercept (0, 2). Parallel to the x-axis, no x-intercept.		
Identity (This is the basic form of a linear equation in the form $y = mx + b$)	$f(x) = x$	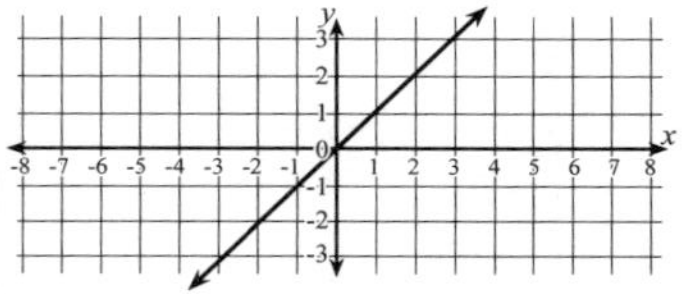	(0, 0)		Linear and Identity function relationship: The Identity Function is the familiar slope intercept linear equation where m = 1 and $b = 0$.
Square or Quadratic	$f(x) = x^2$	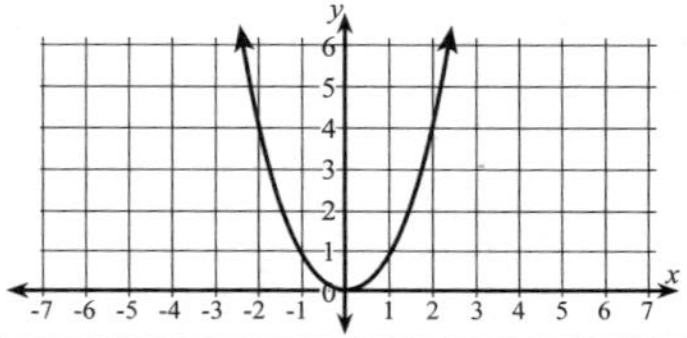	(0, 0)	Min: (0, 0)	
Exponential	$f(x)=b^x; b>0, b\neq1$ ***Note:*** $b=2$ is used for the sketch. If $0 < b < 1$, the graph is reflected over the y-axis.	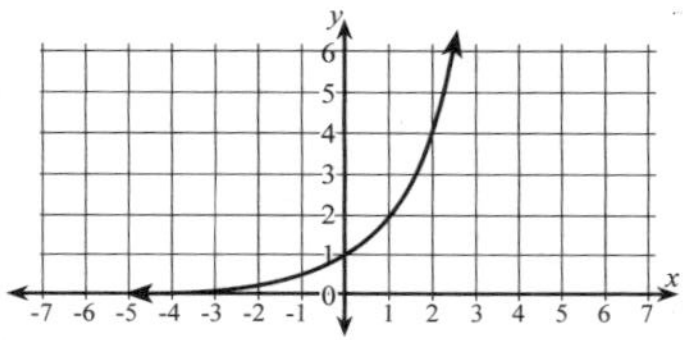	y-intercept at (0, 1). No x-intercept.	The x-axis is an *asymptote* for the graph of this function.	The y-intercept is (0, 1). There is no x-intercept. If $b > 1$, the exponential graph is increasing. If $0 < b < 1$ the graph is decreasing.

Square Root, Cube Root and Absolute Value Parent Functions

Type of $f(x)$	Function Rule	Sketch	Intercepts	Max/Min	Notes
Square Root	$f(x) = \sqrt{x}, x \geq 0$	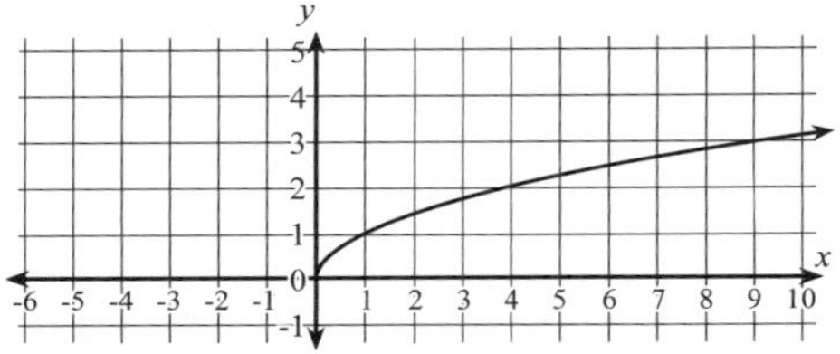	(0, 0)	Minimum (0, 0)	Domain: $x \geq 0$. Negative numbers do not have a real number square root. Range: $y \geq 0$. The $\sqrt{\ }$ symbol indicates the principal root only.
Cube Root	$f(x) = \sqrt[3]{x}$	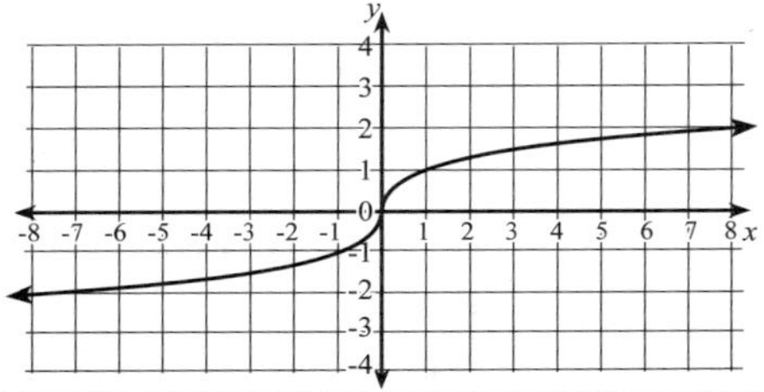	(0, 0)		The domain and range of this function are both all the real numbers.
Absolute Value	$f(x) = \lvert x \rvert$	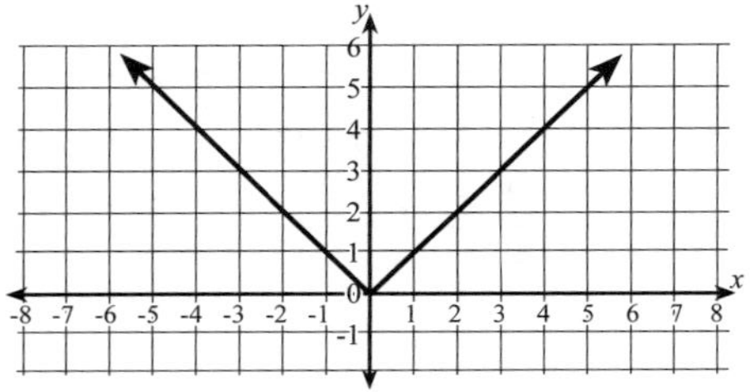	(0, 0)	Minimum (0, 0)	Domain: Real numbers. Range: $y \geq 0$ This function is also considered a piecewise-defined function. $f(x) = \begin{cases} -x \text{ if } x < 0 \\ x \text{ if } x \geq 0 \end{cases}$

4.3

Piecewise-Defined Functions: Piecewise functions have at least two different rules that are used for different parts of their domain. Absolute value functions, step functions, and other functions that are piecewise are shown here. There may be different rules applied to separate parts of the domain.

The rules for the function are written using large brackets with several rules inside. The word "if" is sometimes but not always used. To evaluate a piecewise function for a value of x, determine which part of the domain x is in and use the rule that is applied to that part.

Piecewise Function: This means that from the domain of $-\infty$ to and including 1, the function is $f(x) = x + 2$. The next part of the function is $f(x) = -x$ in the domain 1 to 5 but not including 1 or 5.

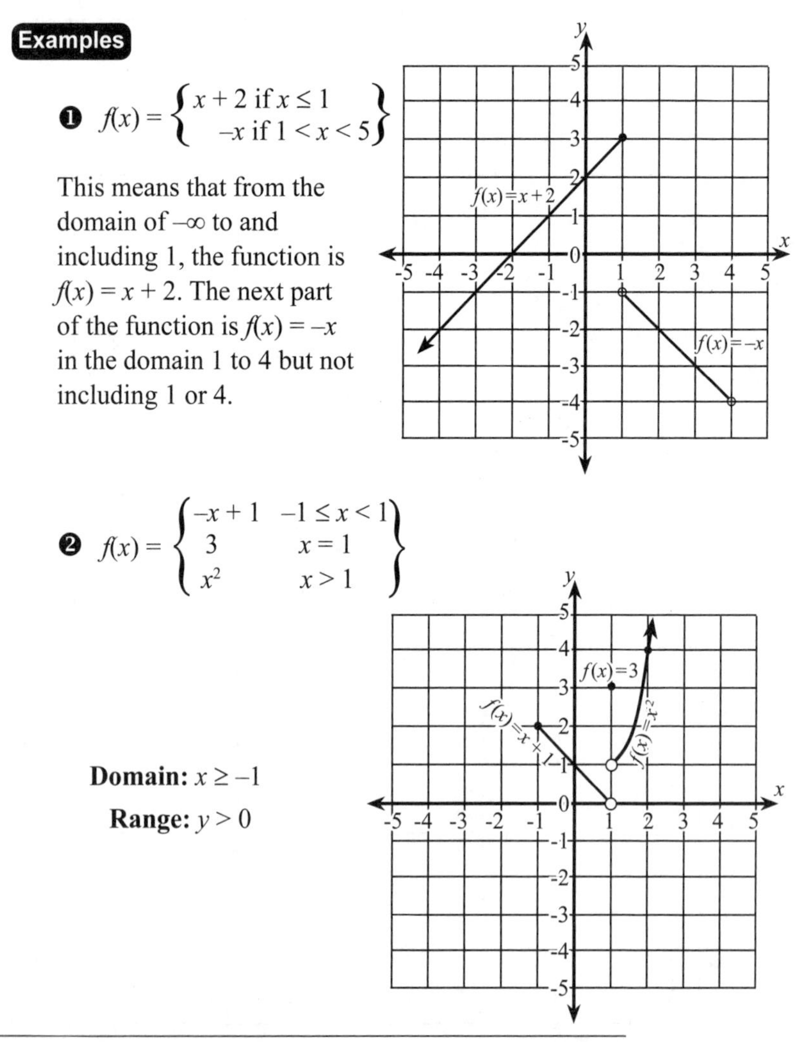

Examples

❶ $f(x) = \begin{cases} x + 2 \text{ if } x \leq 1 \\ -x \text{ if } 1 < x < 5 \end{cases}$

This means that from the domain of $-\infty$ to and including 1, the function is $f(x) = x + 2$. The next part of the function is $f(x) = -x$ in the domain 1 to 4 but not including 1 or 4.

❷ $f(x) = \begin{cases} -x + 1 & -1 \leq x < 1 \\ 3 & x = 1 \\ x^2 & x > 1 \end{cases}$

Domain: $x \geq -1$

Range: $y > 0$

Step Function: Step functions are used for certain situations in which the graph has a constant value over sections of its domain. An example would be if every non-integer value of x down to the next lower integer to obtain $f(x)$. This is called a "greatest integer" function – think of it as the largest or greatest integer less than the given number. The symbol for the greatest integer is $[\![x]\!]$. The graph of a step function looks like a staircase viewed from the side! Sometimes the steps are connected with vertical lines, and sometimes they remain separate from each other.

Step Function: $f(x) = [\![x]\!]$

$$f(x) = \left\{ \begin{array}{r} \ldots\ -2 \text{ if } -2 \leq x < -1 \\ -1 \text{ if } -1 \leq x < 0 \\ 0 \text{ if } 0 \leq x < 1 \\ 1 \text{ if } 1 \leq x < 2\ \ldots \end{array} \right\}$$

The "..." means to continue in the same pattern in the negative direction and in the positive direction.

Domain: Real numbers

Range: Integers

Evaluate:

$f\left(-\frac{1}{3}\right) = -1$

$f(2) = 2$

$f(2.75) = 2$

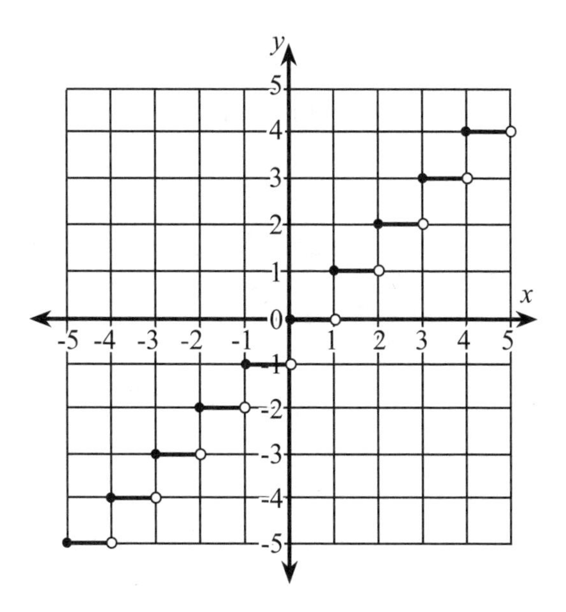

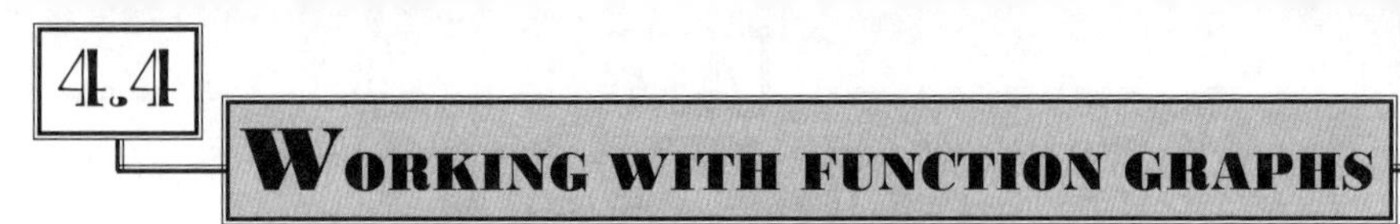

4.4 WORKING WITH FUNCTION GRAPHS

AVERAGE RATE OF CHANGE

Finding the average rate of change over a specific interval of a function is similar to finding the slope of a line connecting two points on a graph. When the graph is linear, we are already familiar with finding the slope, m, using the formula $m = \frac{y_2 - y_1}{x_2 - x_1}$. If the graph is a curve, the points involved can be connected with a line and then the slope of that line is found. The formula for the average rate of change is $\textit{Avg Rate} = \frac{f(x_2) - f(x_1)}{x_2 - x_1}$. This formula can be used with a function given in algebraic form, shown as a graph, or in as a table.

Examples

❶ **Algebraic:** Given the function $f(x) = x^2 + 3x - 5$

a) Find the average rate of change in the interval [2, 10].

Solution: First evaluate $f(2)$ and $f(10)$, then substitute in the formula.

$f(2) = 2^2 + 3(2) - 5 = 4 + 6 - 5 = 5$

$f(10) = 10^2 + 3(10) - 5 = 100 + 30 - 5 = 125$

$$\textit{Avg Rate} = \frac{125-5}{10-2} = \frac{120}{8} = 15$$

This indicates that dependent variable increases (rises) an average of 15 units for each unit of increase of the independent variable in the given interval.

b) Using the same function, find the average rate of change in the interval [–4, –2].

$f(-4) = (-4)^2 + 3(-4) - 5 = 16 - 12 - 5 = -1$

$f(-2) = (-2)^2 + 3(-2) - 5 = 4 - 6 - 5 = -7$

$$\textit{Avg Rate} = \frac{-1-(-7)}{-4-(-2)} = \frac{6}{-2} = -3$$

In the interval between $x = -4$ *and* $x = -2$, the function has an average rate of change of –3. The dependent variable decreases 3 units for each unit of increase in the independent variable.

Note: When graphed, part of this function is increasing and part of this function is decreasing. Therefore the average rate of change may be positive or negative, depending on the interval.

❷ **Graph:** Find the average rate of change in the interval $-3 \le x \le 0$.

Solution: Locate the x values indicated and determine the value of $f(x)$. Substitute.

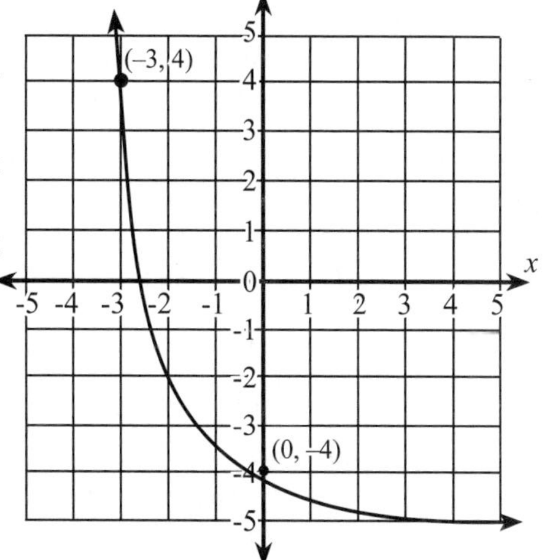

$$Avg\ Rate = \frac{4-(-4)}{-3-0} = \frac{8}{-3}$$

This function's dependent value, $f(x)$ or y, decreases 8 units as the independent variable, x, increases 3 units in the interval indicated.

(A line sketched between the 2 points would have a negative slope.)

Note: This is a decreasing graph so the average rate of change between any 2 different points on the graph would be negative. However, the value of the average rate of change would be different within different intervals. For example, the average rate of change would be less between $x = 2$ and $x = 3$ than it is between $x = -3$ and $x = 0$.

❸ **Table:** Given this table of values for a function

t	$f(t)$
–4	6
–3	1
–2	–2
–1	–3
0	–2
1	1
2	6
3	13
4	22
5	33

Solution: Read the corresponding values of $f(t)$ from the table and substitute.

a) Determine the average rate of change of this function from $t = -3$ to $t = 5$.

$f(-3) = 1$

$f(5) = 33$

$$Avg\ Rate = \frac{33-(1)}{5-(-3)} = \frac{32}{8} = 4$$

b) What is the average rate of change of this function from $t = -1$ to $t = 4$?

$f(-1) = -3$

$f(4) = 22$

$$Avg\ Rate = \frac{22-(-3)}{4-(-1)} = \frac{25}{5} = 5$$

Interpreting Functions

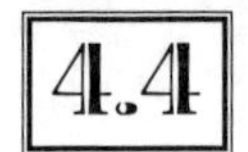

END BEHAVIOR OF FUNCTIONS

End behavior is defined as the trend of the value of $f(x)$ as x approaches ∞ or $-\infty$. End behavior can be determined by looking at the graph of the function. Although it isn't possible to graph all the way to the extremes of the x-axis, the trend of the values of $f(x)$ can be predicted. End behavior does not describe the middle parts of the graphed function.

Terminology used to describe end behavior is varied. The symbol $\rightarrow$ means "approaches." Word descriptions may include upward, rising, or approaching infinity ($\rightarrow \infty$); downward, falling, or approaching negative infinity ($\rightarrow -\infty$). In cases where the graph has an asymptote the end behavior can be described as approaching the asymptote (e.g. Right side $\rightarrow$ positive x-axis).

In a quadratic function or another polynomial function with an even degree (highest exponent is 2, 4, 6,...) the graph has two "arms" – one on the left and one on the right. The end behavior is either both arms rising or both arms falling. The positive or negative value of the coefficient of the x^2 term determines whether upward or downward behavior occurs.
See Figures 1 and 2.

A third degree function, also called a cubic function, has "arms" that go in opposite directions. One is rising upward and the other is falling downward as x changes. The behavior depends on the positive or negative value of the coefficient of the x^3 term. This end behavior also occurs for other functions with an odd degree (3, 5, 7, ...). **See Figures 3 and 4.**

Value of a	Positive	Negative	Positive	Negative
Value of n	Even	Even	Odd	Odd
	Figure 1 $a > 0$	**Figure 2** $a < 0$	**Figure 3** $a > 0$	**Figure 4** $a < 0$
Graph				
End Behavior on *left* side of graph where $x \rightarrow -\infty$	$f(x) \rightarrow \infty$ Graph rises	$f(x) \rightarrow -\infty$ Graph falls	$f(x) \rightarrow -\infty$ Graph falls	$f(x) \rightarrow \infty$ Graph rises
End Behavior on *right* side of graph where $x \rightarrow \infty$	$f(x) \rightarrow \infty$ Graph rises	$f(x) \rightarrow -\infty$ Graph falls	$f(x) \rightarrow \infty$ Graph rises	$f(x) \rightarrow -\infty$ Graph falls

4.4

The graph of an exponential function is in the form $f(x) = n^x$ where n is positive, some special characteristics are present. When $n > 1$ the graph is increasing and approaches the x-axis as x approaches negative infinity. If $0 < n < 1$ the graph is decreasing and approaches the x-axis as x approaches positive infinity. Both have vertical asymptotes that depend on the value of the base of the exponent. **See Figures 5 and 6.**

> **Special Note:** Remember that when an exponent is negative, the base used is the reciprocal of the given base raised to the positive value of the exponent.
>
> The exponential function $f(x) = 5^{-x}$ is equivalent to $f(x)=\left(\frac{1}{5}\right)^x$ thus changing the value of n from $n > 1$ to $0 < n < 1$. That changes the graph from increasing to decreasing.
>
> Likewise if $0 < n < 1$ is the base with a negative value of x, using the reciprocal with a positive exponent makes the base, $n > 1$, and the graph is then increasing. $f(x) = \left(\frac{1}{2}\right)^{-x}$ *or* $f(x) = .5^{-x}$ is equivalent to $f(x) = 2^x$.

<u>End Behavior of Graphs With Asymptotes</u>: Exponential functions when graphed have a line which one arm of the graph approaches but does not cross. This is called an asymptote.

Value of n	$n > 1$	$0 < n < 1$
Graph	**Figure 5** $f(x) = 2^x$	**Figure 6** $f(x) = .5^x$
End Behavior on *left* side of graph where $x \to -\infty$	Graph approaches the negative x-axis. $f(x) \to 0$	Graph is rising, $f(x) \to \infty$
End Behavior on *right* side of graph where $x \to \infty$	Graph is rising, $f(x) \to \infty$	Graph approaches the positive x-axis. $f(x) \to 0$

Interpreting Functions

Note: The descriptions of end behavior do not tell us anything about the middle of the graphs. The examples shown in the tables above are simple functions. **Figures 7 and 8** demonstrate functions that have the same end behaviors as the simple ones in the tables, but the middle of the graphs are quite different.

<ins>**Figure 7**</ins>

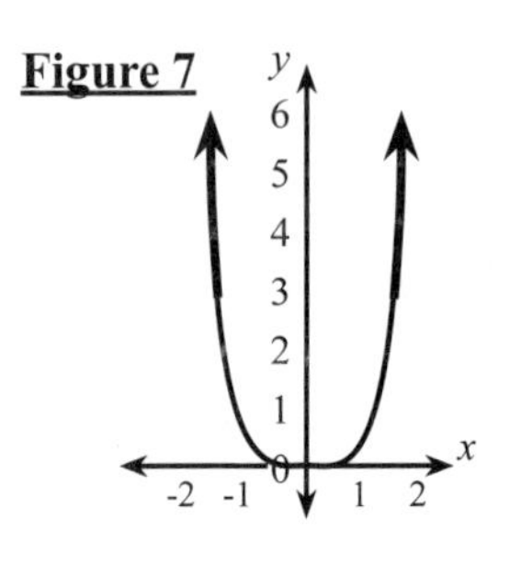

Figure 7 is a sketch of the function $f(x) = x^6$. It is an *even function*, 6th degree, with a *positive coefficient*. The end behavior on both sides is upward. The flat appearance of the middle of the graph is not described.

<ins>**Figure 8**</ins>

Figure 8 is the sketch of the function $f(x) = x^5 - x^4 + x^3 + x^2 - x - 3$. It is an *odd degree function* and has a *positive coefficient*. It follows the expected pattern for odd degree functions – the right arm upward, the left arm downward. The middle of the graph does not change the pattern of end behavior.

<ins>READING FROM A GRAPH OF A FUNCTION</ins>

Locate the x value you are working with and read the corresponding y value on the graph.

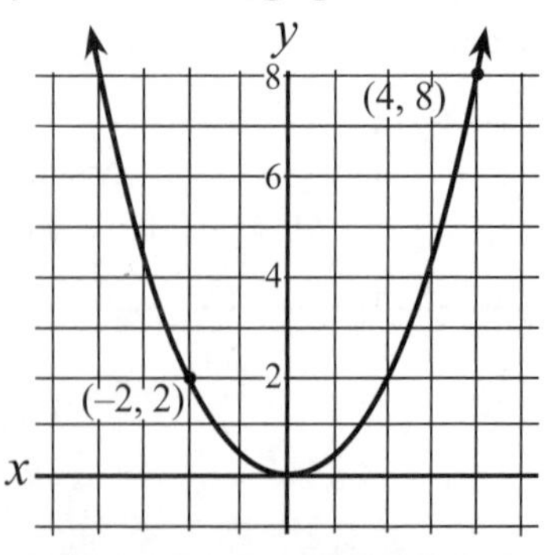

$f(x) = \frac{1}{2}x^2$ is graphed here.

Evaluate $f(-2)$: Locate $x = -2$ and find y above it at 2.

Evaluate $f(4)$: Locate $x = 4$, find y above it at 8.

In this example the function is given. Sometimes just the graph is shown.

FUNCTIONS AND TRANSFORMATIONS

Review of Parent Functions

Identity (Linear): $f(x) = x$

Quadratic: $f(x) = x^2$

Exponential: $f(x) = b^x$

Radical: $f(x) = \sqrt{x}$

Cubic: $f(x) = x^3$

Absolute Value: $f(x) = |x|$

Rules for Transformations of Functions

$f(x)$ represents the parent function

$f(x) = x^2$ and the number 3 is used to demonstrate the rules on this page. For more parent functions and their transformations, see the next 3 pages.

$f(x) + a$: moves the graph up a units.
Ex: $f(x) + 3 = x^2 + 3$

$f(x) - a$: moves the graph down a units.
Ex: $f(x) - 3 = x^2 - 3$

$f(x + a)$: moves it to the left a units.
Ex: $f(x + 3) = (x + 3)^2$

$f(x - a)$: moves it to the right a units.
Ex: $f(x - 3) = (x - 3)^2$

$-f(x)$: reflects the graph over the x-axis.
Ex: $-f(x) = -x^2$

$f(-x)$: reflects the graph over the y-axis.
Ex: $f(x) = (-x)^2$

$a \cdot f(x)$: stretches the graph vertically (or compresses it horizontally) when $a > 1$.
Ex: $f(x) = 3x^2$

$a \cdot f(x)$: compresses the graph vertically (or stretches it horizontally) when $0 < a < 1$.
Ex: $f(x) = 0.5x^2$

EFFECTS OF TRANSFORMATIONS ON PARENT FUNCTION GRAPHS

Parent Function	**Identity (Linear)** $f(x) = x$ $f(x) = mx + b$	**Quadratic** $f(x) = x^2$	**Exponential** $f(x) = b^x; b > 0, b \neq 1$ (*this example,* $b = 2$)	**Square Root** $f(x) = \sqrt{x}, x \geq 0$	**Absolute Value** $f(x) = \lvert x \rvert$
Parent Graph	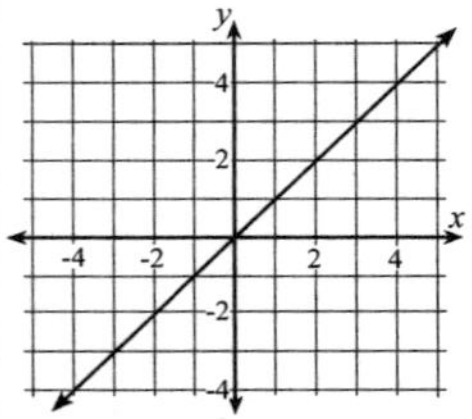	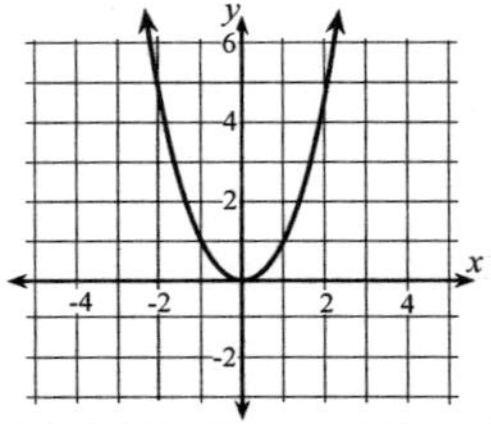	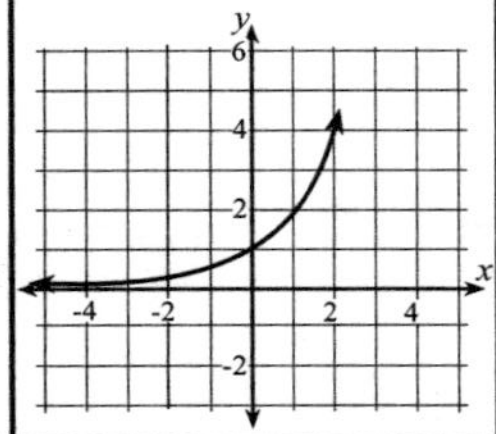	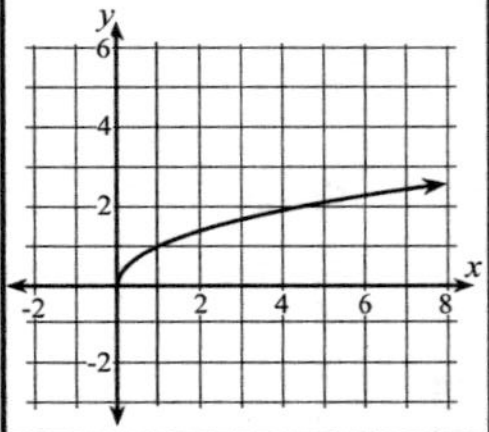	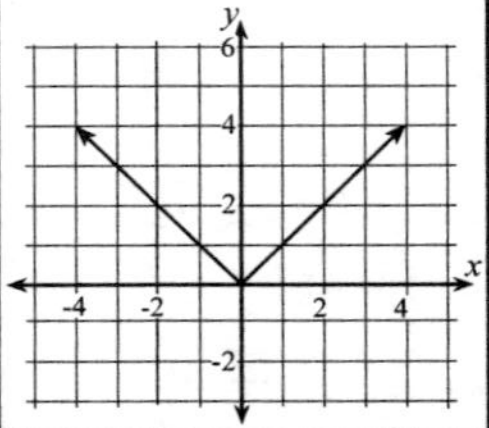
f*(*x*) + *a **Graph moves up *a* units.**	$f(x) = x + 2$ 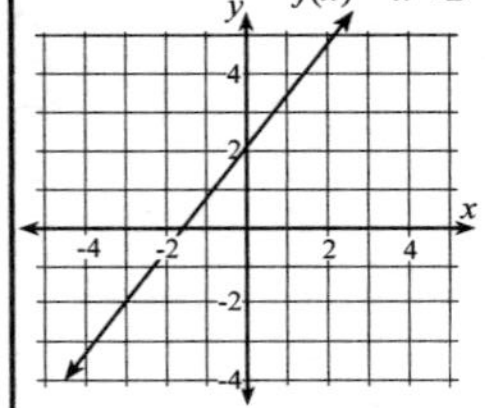	$f(x) = x^2 + 2$ 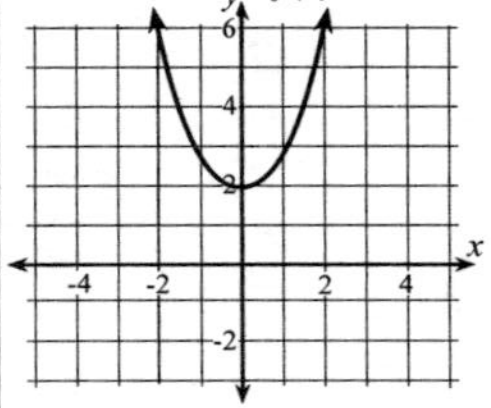	$f(x) = 2^x + 2$ 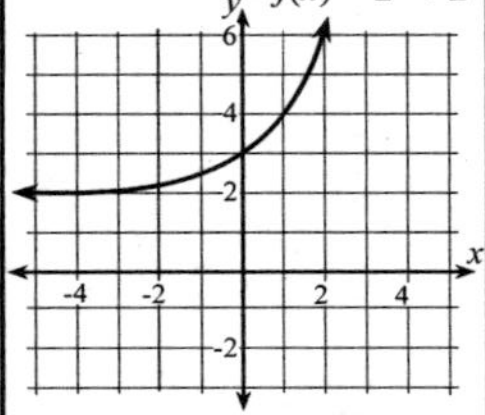	$f(x) = \sqrt{x} + 2$ 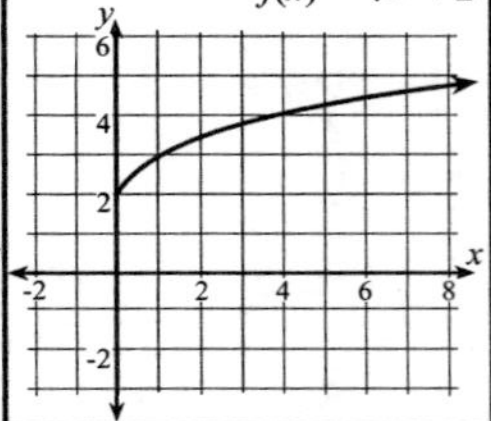	$f(x) = \lvert x \rvert + 2$ 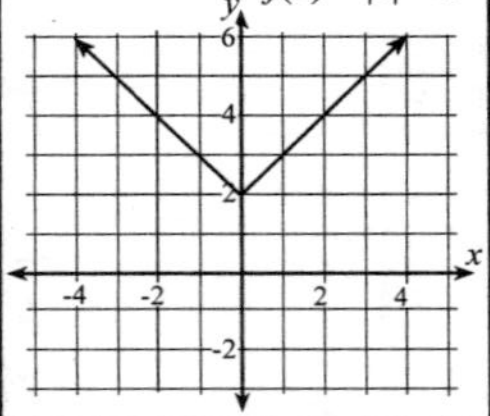
f*(*x*) – *a **Graph moves down *a* units.**	$f(x) = x - 2$ 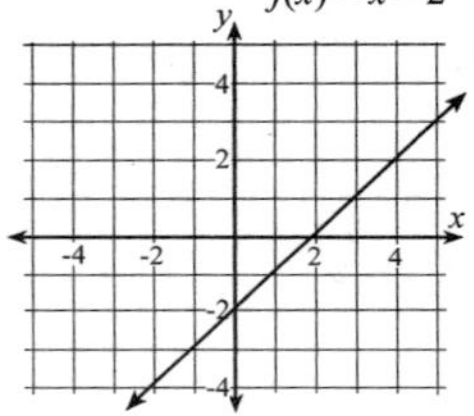	$f(x) = x^2 - 2$ 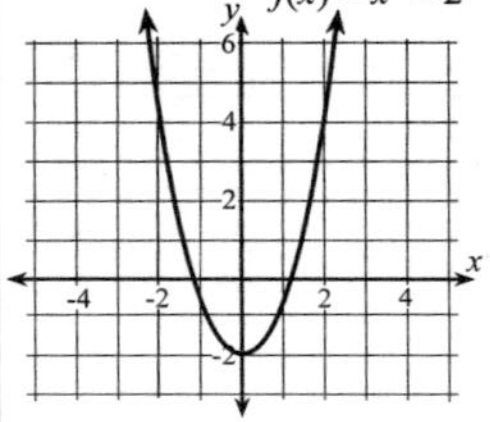	$f(x) = 2^x - 2$ 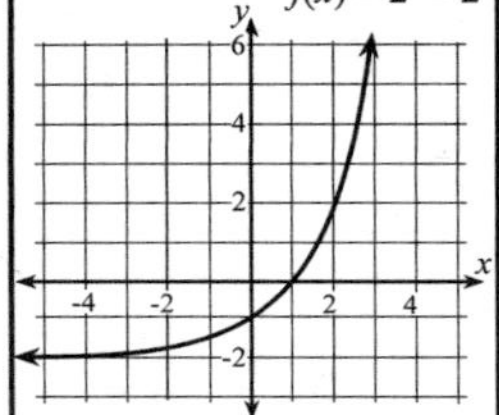	$f(x) = \sqrt{x} - 2$ 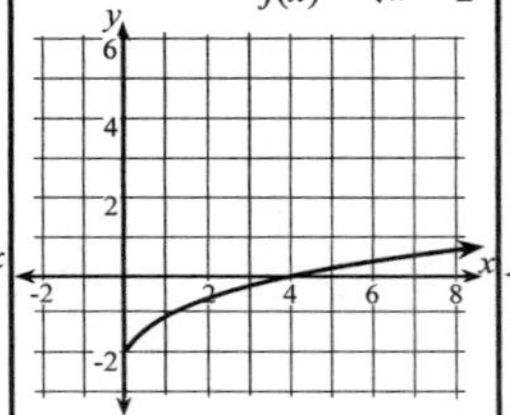	$f(x) = \lvert x \rvert - 2$ 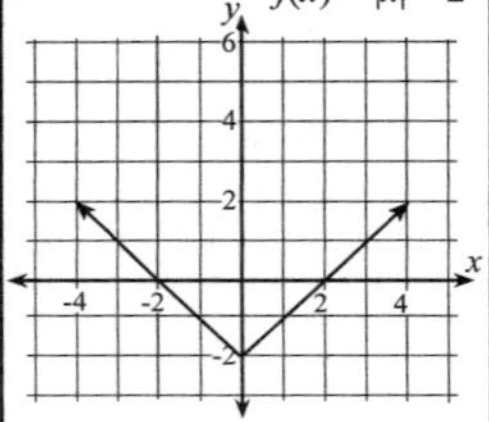

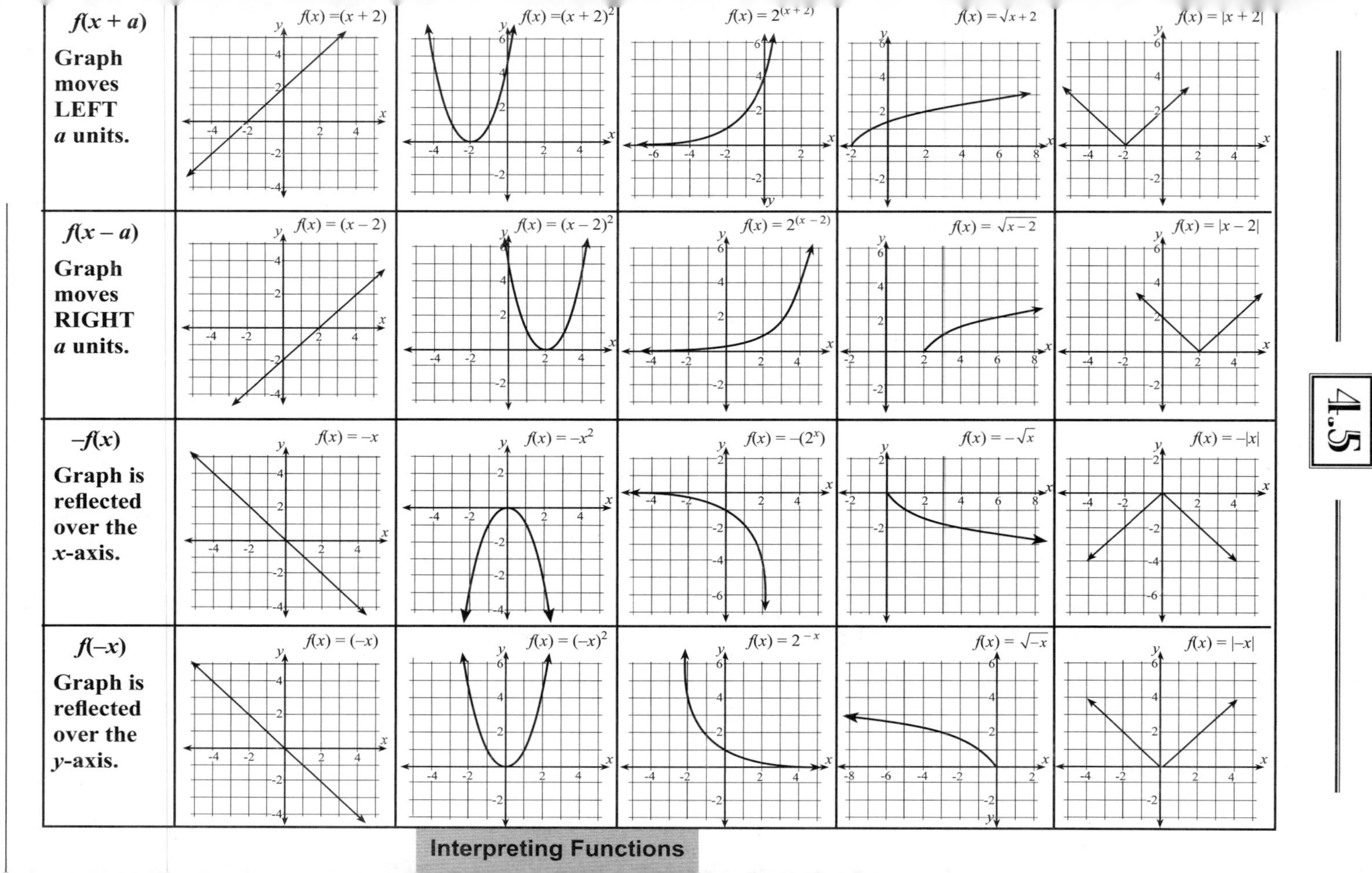

$f(x+a)$
Graph moves LEFT a units.
$f(x) = (x+2)$
$f(x) = (x+2)^2$
$f(x) = 2^{(x+2)}$
$f(x) = \sqrt{x+2}$
$f(x) = |x+2|$
$f(x-a)$
Graph moves RIGHT a units.
$f(x) = (x-2)$
$f(x) = (x-2)^2$
$f(x) = 2^{(x-2)}$
$f(x) = \sqrt{x-2}$
$f(x) = |x-2|$
$-f(x)$
Graph is reflected over the x-axis.
$f(x) = -x$
$f(x) = -x^2$
$f(x) = -(2^x)$
$f(x) = -\sqrt{x}$
$f(x) = -|x|$
$f(-x)$
Graph is reflected over the y-axis.
$f(x) = (-x)$
$f(x) = (-x)^2$
$f(x) = 2^{-x}$
$f(x) = \sqrt{-x}$
$f(x) = |-x|$

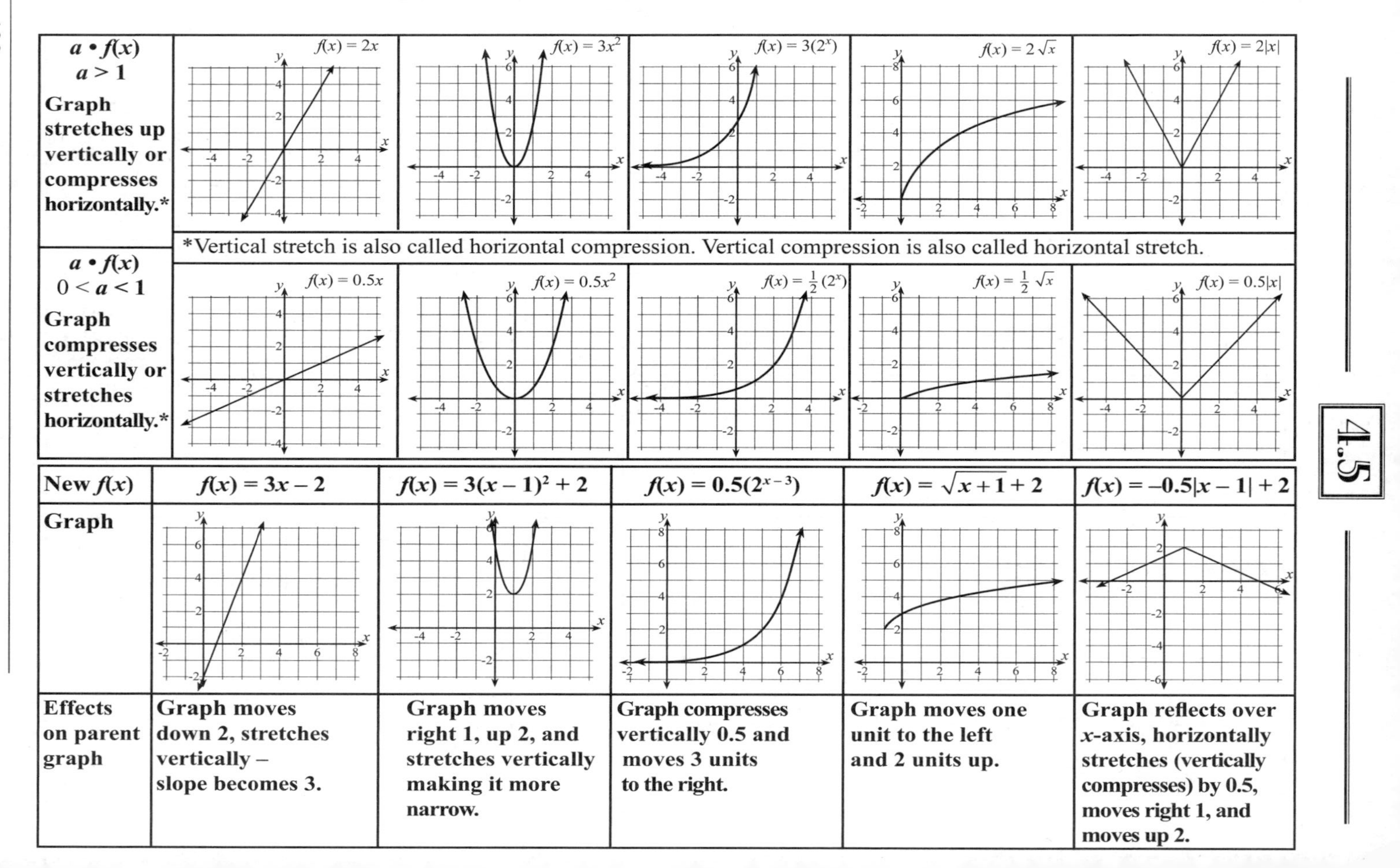

$a \bullet f(x)$, $a > 1$ — Graph stretches up vertically or compresses horizontally.*	$f(x) = 2x$	$f(x) = 3x^2$	$f(x) = 3(2^x)$	$f(x) = 2\sqrt{x}$	$f(x) = 2\|x\|$
*Vertical stretch is also called horizontal compression. Vertical compression is also called horizontal stretch.					
$a \bullet f(x)$, $0 < a < 1$ — Graph compresses vertically or stretches horizontally.*	$f(x) = 0.5x$	$f(x) = 0.5x^2$	$f(x) = \frac{1}{2}(2^x)$	$f(x) = \frac{1}{2}\sqrt{x}$	$f(x) = 0.5\|x\|$

New $f(x)$	$f(x) = 3x - 2$	$f(x) = 3(x - 1)^2 + 2$	$f(x) = 0.5(2^{x-3})$	$f(x) = \sqrt{x+1} + 2$	$f(x) = -0.5\|x - 1\| + 2$
Graph					
Effects on parent graph	Graph moves down 2, stretches vertically – slope becomes 3.	Graph moves right 1, up 2, and stretches vertically making it more narrow.	Graph compresses vertically 0.5 and moves 3 units to the right.	Graph moves one unit to the left and 2 units up.	Graph reflects over x-axis, horizontally stretches (vertically compresses) by 0.5, moves right 1, and moves up 2.

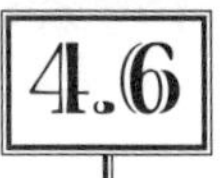

QUADRATIC FUNCTIONS

See also Quadratic Equations – Unit 3

SOLVING QUADRATIC EQUATIONS (FUNCTIONS) GRAPHICALLY

Solving Parabolas: The graph of a quadratic equation (function) in the form $y = ax^2 + bx + c$ where a, b, and c are real numbers and $a \neq 0$ is a parabola. It can also be written as a function: $f(x) = ax^2 + bx + c$. The graph has a "U" shape. The position and exact shape of the "U" are determined by using the values of a, b, and c in the equation.

Up or Down?

If a is negative, the graph of the parabola will open downward, ∩.
If a is positive, the graph opens upward, U.

***Y*-Intercept:** In the equation, $y = ax^2 + bx + c$, the constant, c, is the y-intercept.

Axis of Symmetry: The line of reflection of a parabola. Points on one side of the axis of symmetry are mirror images of the points on the other side. The axis of symmetry goes through the turning point of the parabola. To find the equation of the axis of symmetry use the formula $x = -b/2a$. The axis of symmetry can also sometimes be read directly from a completed graph.

Turning Point or Vertex: The maximum (if parabola opens downward) or minimum (if the parabola opens upward) point on the graph. The turning point is on the axis of symmetry. The x coordinate of the turning point can be found by using $x = -b/2a$. The y coordinate of the turning point can then be found by substituting the value of x into the original equation, $y = ax^2 + bx + c$. The x coordinate of the turning gives a "center" for the table of values used to graph a parabola. This can also be read from the graph at times.

Roots: The roots, or zeros, of the equation are the x values of the points where the parabola crosses the x-axis. There can be one, two, or no real roots. If it doesn't cross the x-axis at all, then there are no real roots. If it just touches the axis, there is one real root, and if it intersects it in two places, then there are two real roots. The roots can be found by reading the graph - accuracy in graphing is essential here. To check the roots: Since $y = 0$ at the roots, substitute 0 for y in the equation and substitute your answers for x. Remember the roots are the solutions of the equation or the value of x when $y = 0$. They are the opposite numbers from the factors used to solve the equation algebraically.

4.6

Solutions: The values of x and y that make the function true. The graph of a function represents all the (x, y) solutions for that function or equation. Write as an ordered pair. (x, y) or $x =$ ____, $y =$ ____.

Maximum/Minimum Values: The y value of the vertex is either the maximum point of the graph if a is negative or the minimum point if a is positive. These are also called maxima and minima.

Finding the Domain: Remember that the domain is always the real numbers unless specified otherwise or restricted by the contents of the relation or function because of a fraction or radical or both! It is the largest set of numbers that will result in a real number solution.

To find restrictions, determine what values of x will not give a real number output. (Since the "default" set is all the real numbers, sometimes just the restrictions are written to indicate the domain.)

Examples

❶ 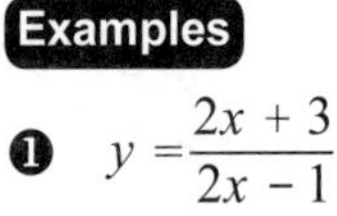

$y = \dfrac{2x+3}{2x-1}$

$2x - 1 \neq 0$

$x \neq \dfrac{1}{2}$

In a fraction the denominator cannot equal zero. That restricts the domain.

Domain: $\{\mathrm{R}, x \neq \frac{1}{2}\}$

❷ 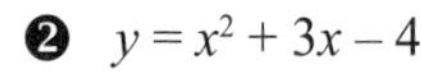

$y = x^2 + 3x - 4$

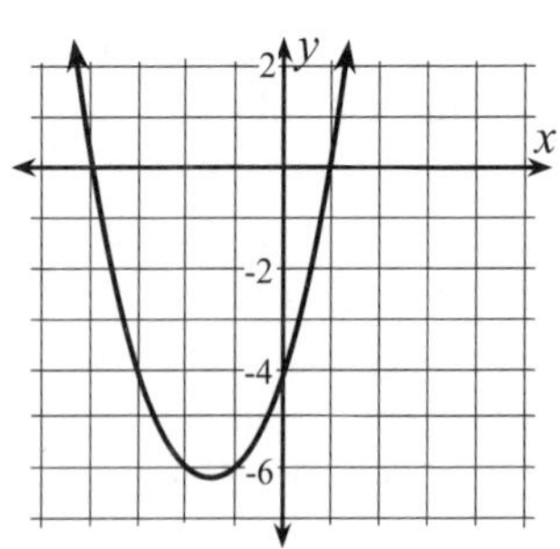

Domain: No restrictions. The domain is the set of Real numbers.

Range: The minimum point on the graph of this equation is $(-1.5, -6.25)$ The range is $\{y : y \geq -6.25\}$

It is a function because the vertical line test works.

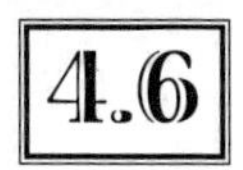

4.6

Solve Graphically: Solving quadratic equation graphically gives not only the roots, but all the possible values of (x, y) that make the equation true. ***Note:*** This work can be done nicely using a graphing calculator. Your teacher will advise you as to what work you will need to show on your paper when using a graphing calculator.

Steps

1) Solve the quadratic equation for y in terms of x. It will be in the form $y = ax^2 + bx + c$, or written in function notation as $f(x) = ax^2 + bx + c$. Make note of the values of a, b, and c.
2) Find the axis of symmetry: $x = -b/2a$. This also provides the x value of the turning point.
3) Make a table of values for the parabola. Use at least three integral values on each side of the x coordinate of the turning point. (Try not to use fractions or decimals for the values of x you choose.) Sometimes the interval of the values for x to be used are given.
4) Substitute the values of x to find the y-coordinate of each point. SHOW SUBSTITUTIONS.
5) Plot the points and sketch the graph accurately and with a smooth curve.
6) LABEL the vertex and two more points on the curve – one on each side of the vertex. Label the roots if possible. Otherwise label one point on each side of the vertex. Label the curve with the original equation.

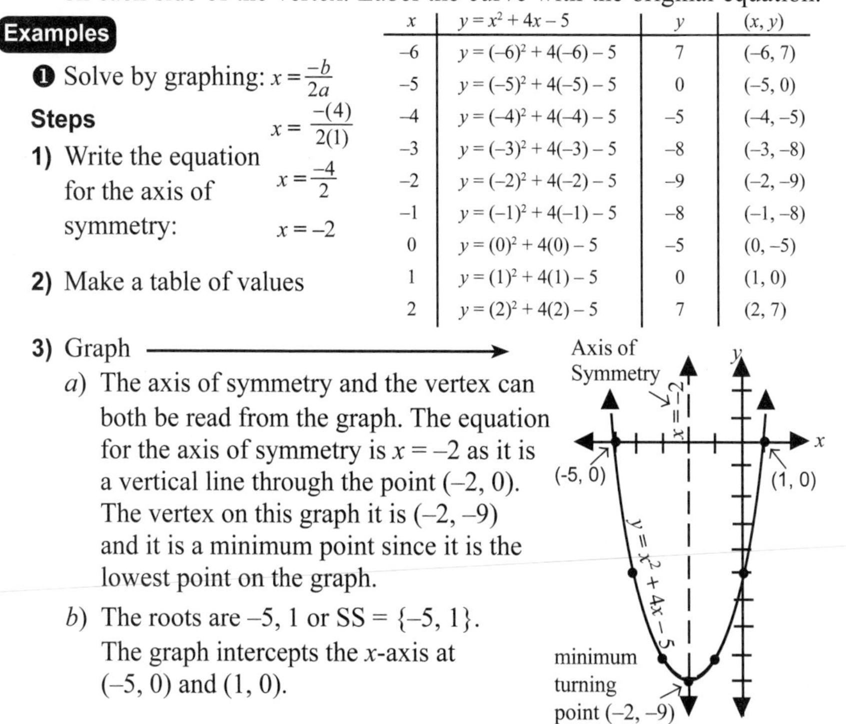

Examples

❶ Solve by graphing: $x = \frac{-b}{2a}$

Steps

1) Write the equation for the axis of symmetry:

$$x = \frac{-(4)}{2(1)}$$

$$x = \frac{-4}{2}$$

$$x = -2$$

2) Make a table of values

x	$y = x^2 + 4x - 5$	y	(x, y)
–6	$y = (-6)^2 + 4(-6) - 5$	7	(–6, 7)
–5	$y = (-5)^2 + 4(-5) - 5$	0	(–5, 0)
–4	$y = (-4)^2 + 4(-4) - 5$	–5	(–4, –5)
–3	$y = (-3)^2 + 4(-3) - 5$	–8	(–3, –8)
–2	$y = (-2)^2 + 4(-2) - 5$	–9	(–2, –9)
–1	$y = (-1)^2 + 4(-1) - 5$	–8	(–1, –8)
0	$y = (0)^2 + 4(0) - 5$	–5	(0, –5)
1	$y = (1)^2 + 4(1) - 5$	0	(1, 0)
2	$y = (2)^2 + 4(2) - 5$	7	(2, 7)

3) Graph ⟶

   a) The axis of symmetry and the vertex can both be read from the graph. The equation for the axis of symmetry is $x = -2$ as it is a vertical line through the point (–2, 0). The vertex on this graph it is (–2, –9) and it is a minimum point since it is the lowest point on the graph.
   b) The roots are –5, 1 or SS = {–5, 1}. The graph intercepts the x-axis at (–5, 0) and (1, 0).

❷ $2x^2 = y$

$a = 2,\ b = 0,\ c = 0$

$x = \frac{-b}{2(a)} = \frac{0}{2(2)} = 0$ *Axis of symmetry* : $x = 0$

x values of vertex = 0

Table of Values		
x	$2x^2$	y
–3	$2(-3)^2$	18
–2	$2(-2)^2$	8
–1	$2(-1)^2$	2
0	$2(0)^2$	0
1	$2(1)^2$	2
2	$2(2)^2$	8
3	$2(3)^2$	18

Graph:

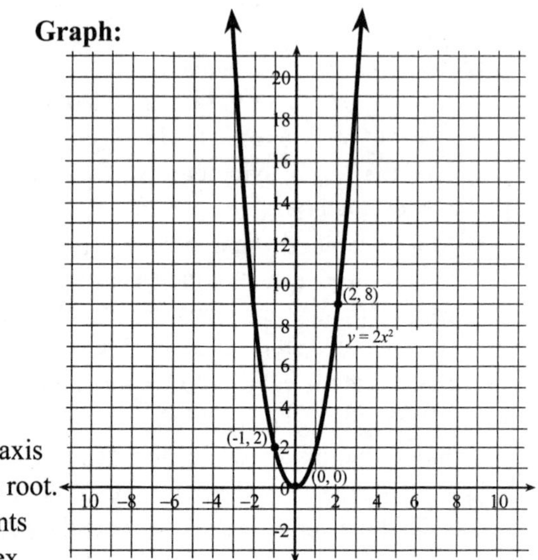

This graph is tangent to the x-axis at $(0, 0)$. The equation has one root. Be sure to label two more points – one on each side of the vertex.

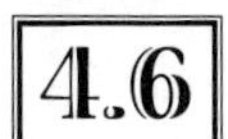

APPLICATIONS WITH QUADRATIC FUNCTIONS

Quadratic Word Problems: Word problems that use the words "squared" or "the product of" often result in equations which contain a variable to the 2nd power: x^2. The let statement contains only one variable. When the equation is written, it may require multiplication of the parts of the let statement. When the multiplication is performed, the variable will have an exponent. Quadratic word problems are usually solved algebraically but could be done graphically as well.

Examples

❶ Jack had to get three more pails of water than Jill did. The product of the number of pails they had to get was 40. How many pails of water did each have to get?

Let x = Number of pails of water Jill had to get

$\therefore x + 3$ = the number of pails of water Jack had to get

$x(x + 3) = 40$

$x^2 + 3x = 40$

$x^2 + 3x - 40 = 0$

$(x - 5)(x + 8) = 0$

$x - 5 = 0$; $x + 8 = 0$

$\boxed{x = 5}$; $x = -8$ reject

It isn't possible to carry –8 pails of water, so the solution –8 is rejected. Use 5 as the only correct answer for x.

Jack's pails: $x + 3 = (5) + 3$

$\boxed{x + 3 = 8}$

Conclusion: Jill had to get 5 pails and Jack had to get 8 pails of water.

❷ A garden is being planned for the front of the school. The width is 6 feet shorter than the length. A local landscaping business offered to donate 135 square feet of mulch to cover the garden. What should the dimensions of the garden be to make use of all the donated mulch?

Let x = *length*

$\therefore x - 6$ = *width*

Formula: $A = lw$

$135 = x(x - 6)$

$135 = x^2 - 6x$

$x^2 - 6x - 135 = 0$

$(x - 15)(x + 9) = 0$

$x - 15 = 0 \quad x + 9 = 0$

$x = 15 \quad x = -9$ *reject* (Can't be negative feet.)

$x - 6 = 9$

Conclusion: The garden can be 15 feet long and 9 feet wide.

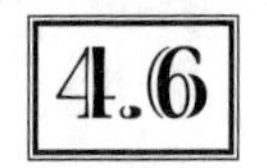

Writing the Equation of a Parabola: The roots of a parabola drawn on a graph can be used to find an equation of the parabola it represents if the coefficient of the x term is 1. Work backwards using factors. In the example below, the roots are –5 and 1. The factors will be formed using the numbers opposite them so the factors will have +5 and –1 in them.

Steps

1) Develop two simple equations that will result in –5 and 1 when solved. The numbers will be opposite those found as the roots or solutions. $x + 5 = 0$; $x - 1 = 0$
2) Working backward, make the equations into two factors = 0. $(x + 5)(x - 1) = 0$
3) Multiply. $x^2 + 4x - 5 = 0$
4) Since $y = 0$ only at the roots, we can exchange the 0 for y to represent the entire parabola. $x^2 + 4x - 5 = y$

Vertex Form and Standard Form: Sometimes the roots of the equation are not readily available. The vertex form of an equation can be used to write a quadratic equation from a graph or from the standard form of the quadratic equation.

Standard Form: $y = ax^2 + bx + c$

Vertex Form: $y = (x - h)^2 + k$ where the vertex is at (h, k).

Working with the vertex form and standard form:

Write in Vertex Form:

- If the coordinates of the vertex are known (either given or read from a graph), this is a very simple process. Substitute the values of h and k in the appropriate places in the format shown above for the vertex form.
- If asked to write an equation in vertex form from an equation in standard form, my students prefer doing it in several steps that are already familiar:

Steps

1) Find the x value of the vertex using the formula $x = \frac{-b}{2a}$. This is the formula for the axis of symmetry. Since the vertex is on the axis of symmetry, its x value is the same as the axis of symmetry.
2) Find the y value of the vertex by substituting the x value in the original equation.
3) Write the vertex form as shown above.

Write in Standard Form:

When given an equation in vertex form that needs to be changed to standard form:

- Expand the binomial term by squaring it.
- Use the distributive property to multiply by "a"
- Simplify.

Examples

❶ Write the vertex form of the equation for this parabola and write the equation in standard form.

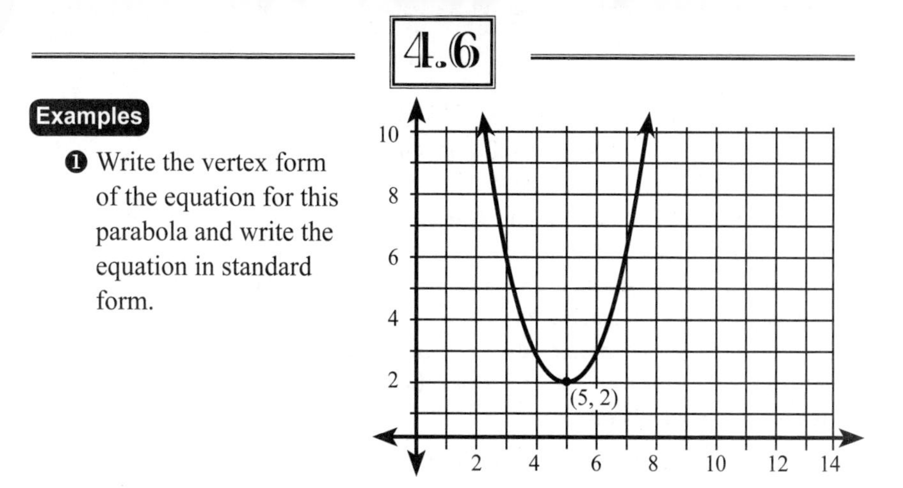

Solution: Read the vertex from the graph. Since the vertex is (5, 2), then $h = 5$ and $k = 2$. In this example, $a = 1$. (Notice that this equation has no roots or zeros – it doesn't intersect the x-axis.)

Vertex Form: $y = (x - 5)^2 + 2$

Standard Form: $y = (x - 5)^2 + 2$

$y = (x^2 - 10x + 25) + 2$ — Square the binomial. Since $a = 1$, there is no need to distribute.

$y = x^2 - 10x + 27$ — Simplify.

❷ Write the vertex form of this quadratic equation: $x^2 + 10x + 33 = y$

Steps

1) Find the x value, h, of the vertex: $x = \dfrac{-b}{2a} = \dfrac{-10}{2} = -5$

2) Find the y value, k, of the vertex: $y = (-5)^2 + 10(-5) + 33 = 8$

3) Substitute in the format needed for a vertex form: $y = (x - h)^2 + k$

$y = (x - (-5))^2 + 8$

$y = (x + 5)^2 + 8$

❸ Write the following equation in vertex form: $y = 2x^2 + 36x + 170$

$y = 2x^2 + 36x + 170$ Notice that $a = 2$.

$x = \dfrac{-36}{2(2)} = -9$ *This is h.*

$y = 2(-9)^2 + 36(-9) + 170$

$y = 8$ *This is k.*

$y = 2(x - (-9))^2 + 8$

$y = 2(x + 9)^2 + 8$

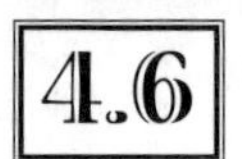

COMPARING QUADRATICS GRAPHICALLY AND ALGEBRAICALLY

It is important to realize that the information obtained from an algebraic solution of a quadratic equation matches the information that is available in a graphic solution. The roots of the equation, also called the zeros, are the x-values of the x-intercept on a graph. The maximum or minimum value of the equation is the vertex of the graph, also called the turning point. Both the x and y values of the vertex can be used to analyze the problem. The algebraic solutions can be done with any of the methods already described. When graphing, the approximate solutions can be read from the graph and obtained using the appropriate functions in the graphing calculator.

Example Two friends were swimming with their new diving equipment and decided to have a diving contest. The contest would involve two measurements. The first would be the depth of the underwater dive, and the second would be the distance traveled underwater measured at the surface.

This equation represents the path of Frank's dive: $f(x) = x^2 - 19x + 70$.
This equation represents the path of George's dive: $g(x) = x^2 - 14x + 33$.

Solve one equation graphically with an appropriate sketch and the other equation using algebraic methods. Determine (and label where appropriate) the vertices and the roots of each equation.

Which boy dove the deepest and by how much? Which boy swam the longest distance underwater and by how much? Use feet for the measuring unit.

Analysis: Choose an equation that looks easy to factor and solve that one algebraically. Solve the other graphically. The water surface is at $y = 0$. Then compare the results.

Algebraic Solution: George's Dive: $g(x) = x^2 - 14x + 33$

$x^2 - 14x + 33 = 0$ Set the function = to zero and factor to find the roots.

$(x - 3)(x - 11) = 0$

$x - 3 = 0 \quad x - 11 = 0$

$x = 3 \qquad x = 11$ The roots of the equation are 3 and 11.

Find the vertex: $x = \frac{-b}{2a} = \frac{-(-14)}{2} = 7$ is the x value of the vertex.

Substitute the value of x in the actual equation to find the y value of the vertex. $g(7) = 7^2 - 14(7) + 33 = -16$

Vertex: $(7, -16)$

Summary: George enters the water at the point $(3, 0)$ and comes back to the surface at $(11, 0)$ so he is underwater for 8 feet measured along the surface. The depth of the dive equals the y value of the vertex. His dive is 16 feet deep.

4.6

Graphic Solution: Frank's Dive: $f(x) = x^2 - 19x + 70$
The table of values, vertex, and roots as well as the graph itself can be found using the graphing calculator. The graph will need to be sketched on the graph paper, so choose several points from the table of values and graph those to create a graph that is fairly accurate. The vertex and both roots must be labeled. The calculator screens are shown here and the paper graph is shown.

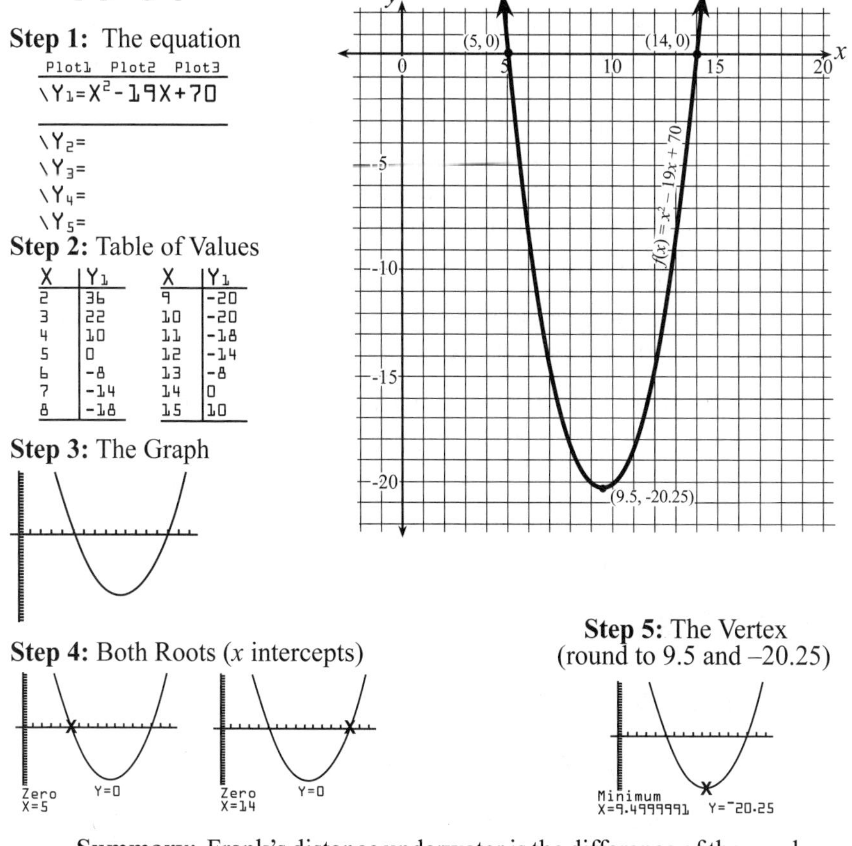

Summary: Frank's distance underwater is the difference of the x-values of the x intercepts: $14 - 5$ or 9 feet measured along the surface of the water. The depth of his dive, the y-value of the vertex, is 20.25 feet.

Conclusion: Compare Frank's and George's dives:
Frank's dive was 20.25 feet deep and George's was only 16 feet. Therefore, Frank dove 4.25 feet deeper than George. The distance Frank was underwater was 9 feet while George was underwater for 8 feet measured along the surface. Therefore, Frank's distance underwater measured along the surface of the water was one foot further than George. Frank's dive was deeper, and longer.

Performing Operations with Functions

Functions are added, subtracted, multiplied and divided using the following notation.

RULES
1) $(f+g)(x)$ means add $f(x)$ and $g(x)$
2) $(f-g)(x)$ means subtract $g(x)$ from $f(x)$
3) $(f \bullet g)(x)$ means multiply $f(x)$ and $g(x)$. Be careful not to confuse this with $f \circ g(x)$ which indicates a composition.
4) $\left(\frac{f}{g}\right)(x)$ means divide $f(x)$ by $g(x)$

Substitute the function itself in place of $f(x)$ or $g(x)$. If given a number or a different variable expression to use, substitute that in place of x in each function and then perform the indicated operations.

When finding the value of the sum, difference, product, or quotient of two functions, two different methods are correct:

- Perform the operation on the functions first, then substitute the given value and evaluate.

or

- Substitute the given value in each function, evaluate then perform the operation. These functions are the ones we'll use for examples.

Examples Given: $f(x) = x^2 + x - 2$ and $g(x) = 3x$

❶ **Addition:** $(f+g)(x) = f(x) + g(x)$
$= x^2 + x - 2 + 3x$
$(f+g)(x) = x^2 + 4x - 2$

Addition with Substitution

- **Substitute, then add:**

Find: $(f+g)(5)$
$[(5)^2 + 5 - 2] + 3(5) = 43$
$(f+g)(5) = 43$

- **Add, then substitute:**

Find: $(f+g)(x+4)$
$f(x) + g(x) = x^2 + x - 2 + 3x = x^2 + 4x - 2$
$(x+4)^2 + 4(x+4) - 2 = x^2 + 8x + 16 + 4x + 16 - 2$
$(f+g)(x+4) = x^2 + 12x + 30$

❷ **Subtraction**: $(f-g)(x) = f(x) - g(x)$

$$= x^2 + x - 2 - 3x$$
$$= x^2 - 2x - 2$$

Subtraction with Substitution

- **Substitute, then subtract:**

Find: $(f-g)(5)$

$((5)^2 + 5 - 2) - (3(5)) = 13$

$(f-g)(5) = 13$

- **Subtract, then substitute:**

Find: $(f-g)(x+4)$

$f(x) - g(x) = x^2 + x - 2 - 3x = x^2 - 2x - 2$

$(x+4)^2 - 2(x+4) - 2 = x^2 + 8x + 16 - 2x - 8 - 2$

$(f-g)(x+4) = x^2 + 6x + 6$

❸ **Multiplication**: $(f \cdot g)(x) = (f(x)) \cdot (g(x))$

$$= (x^2 + x - 2)(3x)$$
$$(f \cdot g)(x) = 3x^3 + 3x^2 - 6x$$

Multiplication with Substitution

- **Substitute, then multiply:**

Find: $(f \cdot g)(5)$

$(5^2 + 5 - 2)(3 \cdot 5) = (28)(15) = 420$

$(f \cdot g)(5) = 420$

- **Multiply, then substitute:**

Find: $(f \cdot g)(x+4)$

$f(x) \cdot g(x) = (x^2 + x - 2)(3x) = 3x^3 + 3x^2 - 6x$

$3(x+4)^3 + 3(x+4)^2 - 6(x+4)$

$3(x^3 + 12x^2 + 48x + 64) + 3(x^2 + 8x + 16) - 6x - 24$

$3x^3 + 36x^2 + 144x + 192 + 3x^2 + 24x + 48 - 6x - 24$

$(f \cdot g)(x+4) = 3x^3 + 39x^2 + 162x + 216$

4.8

WORKING WITH FUNCTIONS

COMPARISON OF A LINEAR, A QUADRATIC AND AN EXPONENTIAL FUNCTION USING A TABLE OF VALUES

Linear functions (**Figure 1**) grow by equal differences over equal intervals. Exponential functions (**Figure 2**) grow by equal factors over equal intervals. Quadratic functions (**Figure 2**) grow by multiplying the difference between equal intervals of the $f(x)$ values of the function by one factor. The x value increases by one unit each time, but the value of $f(x)$ changes depending on the type of function.

Examples

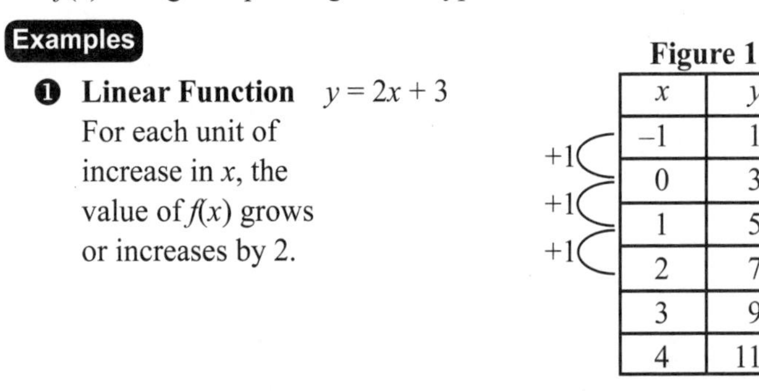

❶ **Linear Function** $y = 2x + 3$
For each unit of increase in x, the value of $f(x)$ grows or increases by 2.

Figure 1

x	y
–1	1
0	3
1	5
2	7
3	9
4	11

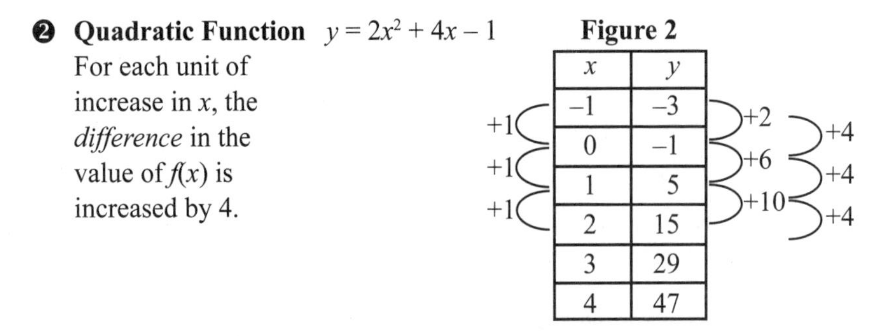

❷ **Quadratic Function** $y = 2x^2 + 4x - 1$
For each unit of increase in x, the *difference* in the value of $f(x)$ is increased by 4.

Figure 2

x	y
–1	–3
0	–1
1	5
2	15
3	29
4	47

❸ **Exponential Function** $y = 2^x$
For each unit of increase in x, the value of $f(x)$ increases by a factor of 2.

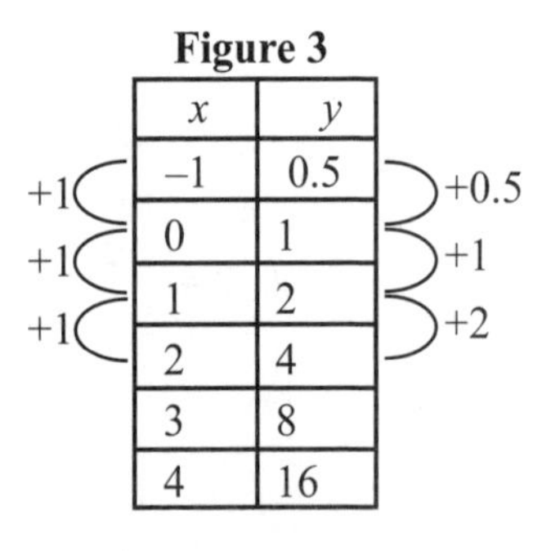

Figure 3

x	y
–1	0.5
0	1
1	2
2	4
3	8
4	16

Summary: You can see by the tables on the previous page that in each example the functions are growing over an equal interval. On the left, the difference in the input is always one unit.

Linear functions grow by equal differences over these equal intervals. On the right of the linear equation table, the difference of the outputs is always an equal amount.

Quadratic functions will always grow by equal differences after two calculations are made. In the example, the original outputs differ by 2, 6, and 10. When the second difference is taken, the result is always equal, in this case four.

Exponential functions grow, not by equal differences, but by equal factors over equal intervals. When finding the differences in the outputs, the result will always be an equal factor.

FUNCTIONS AND SEQUENCES

Arithmetic Sequence: An arithmetic sequence (also called an arithmetic progression) is a linear function with the domain being the set of natural numbers {1, 2, 3, …}. Each term is calculated from the prior term by adding or subtracting some constant number called the common difference.

Examples

❶ {1, 6, 11, 16, 21, 26, …} add the constant 5 to get the next term in the sequence. The common difference is 5.

❷ Marge keeps adding friends to her social networking account. She began with only 4 friends, and then each subsequent week had 22, 40, 58, and then 76 friends. If the pattern continues, how many friends will she have in 4 more weeks? Write a function to represent this sequence.

Solution: Based the difference between subsequent weeks, Marge is adding 18 friends each week. In four more weeks she will have $76 + 4(18) = 148$ friends.
$f(x) = 4 + 18x$ where x is the number of weeks.

❸ Write a function to represent this graph of an arithmetic sequence. What is the value of the common difference?

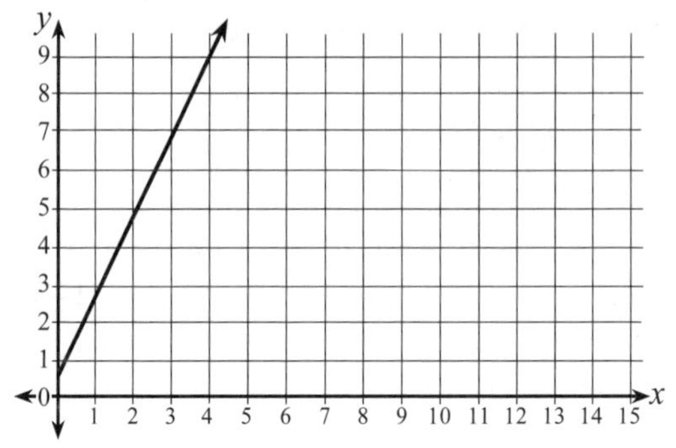

Solution: The initial value here is 1 (the y intercept). Subsequent points are (1, 3), (2, 5), (3, 7).... For each increase of 1 unit in the x value, y increases by 2. The common difference is 2. The function is $f(x) = 2x + 1$.

Geometric Sequence: A geometric sequence (also called a geometric progression) occurs when subsequent terms are found by multiplying by the same constant number. When this constant, called the common ratio, is a number not equal to –1, 0, or 1, the geometric sequence is an exponential function. The constant number is called the geometric ratio and can be found by dividing a number in the sequence by the preceding number. The same ratio is applied to each pair of numbers in the sequence.

Examples

❶ {3, 6, 12, 24, 48, 96, …} multiply by 2 to get the next term in the sequence. The common ratio is 2. The function is $f(x) = 3(2^x)$.

❷ Sales in the ice cream business have tripled for Jennifer in each of the last three months. (it was summer!) During the last month, she made $400 profit. If the pattern continues, what will be her total profit for the next three months? Write a function to represent this situation.

Solution: The common ratio here is 3. Each subsequent month's profit is multiplied by 3. The initial value as given in the problem is $400.

Month 1 = 3($400) = $1200

Month 2 = 3($1200) = $3600

Month 3 = 3($3600) = $10,800

Total profit = $15,500

$f(x) = 400\ (3^x)$ where x is the number of months.

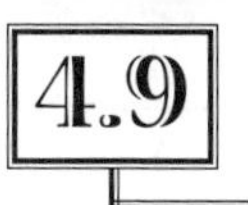

CREATING FUNCTIONS

Creating a Function From a Graph

Examples

❶ Jeff graphed his wages and tips after several weeks of delivering newspapers. Given the graph, write his earnings as a function of the number of newspapers that he delivered.

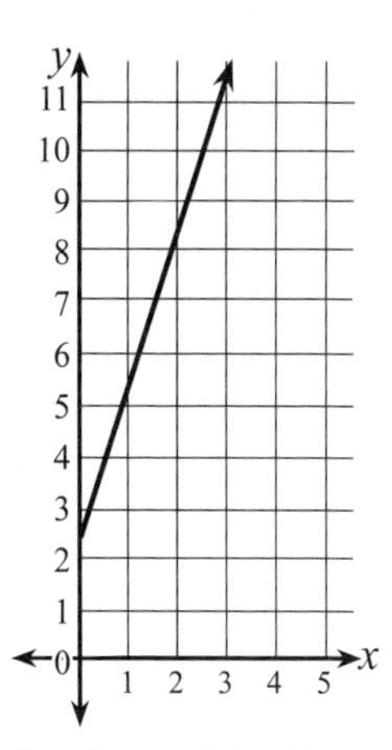

Solution: This is the graph of a *linear* function. The domain is $[0, \infty)$ since he will not sell less than zero newspapers.

Determine the y-intercept from the graph: $y = 2.5$

Choose two points on the graph and determine the slope of the line.

$(0, 2.5)$ and $(1, 5.5)$

$$m = \frac{5.5 - 2.5}{1 - 0} = \frac{3}{1} = 3$$

Write the function in slope-intercept form: $f(x)$ or $y = mx + b$

$$f(x) = 3x + 2.5$$

 Write a function to represent this graph. Identity the type of function and describe the values used to write the function.

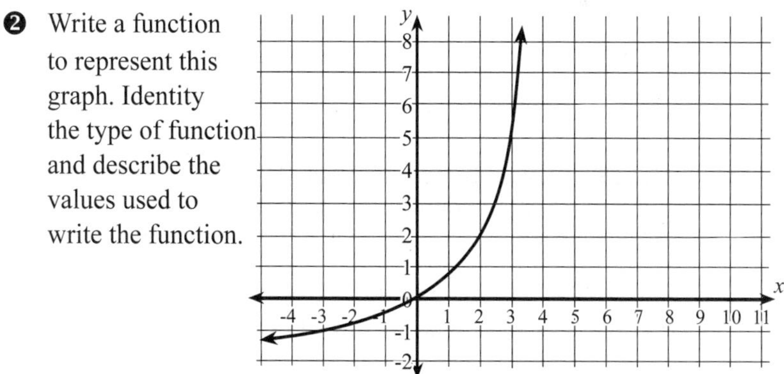

Solution: This is an *exponential* function. The parent exponential function goes through the point $(0, 1)$. This graph goes through $(0, 0)$ so it has been shifted down one unit. To determine the parent function, move it up mentally (or make a sketch) to determine the value of the common ratio easily. The parent points would be $(1, 2)$, $(2, 4)$, $(3, 8)$ … For each unit that x increases, y is multiplied by 2. The common ratio is 2. The function is: $f(x) = 2^x - 1$.

4.9

Creating a Function From a Description of a Relationship

Examples

❶ Suzie makes pillowcases to sell at a monthly craft fair in her town. Each month she spends \$30 on fabric and sewing materials from the local craft store. At the fair, she sells each completed pillowcase for a total of \$10, including tax. Express Suzie's profit as a function of the number of pillowcases that she sells in one month.

Solution: Suzie's monthly spending of \$30 will create a decrease in her profit each month. Each item she sells provides \$10 of income. For each additional pillowcase she sells, her profit increases by \$10. This is a *linear* function. The domain is $0 \leq x < \infty$ since she cannot sell less than zero pillowcases.

$$f(x) = 10x - 30$$

 Jane is raising rabbits for a summer project. She has one pair of rabbits to begin with. Each week the number of pairs doubles. Write a function to represent Jane's summer project based on the number of weeks

Solution: Each week the number of rabbits doubles so 2 is the base for this *exponential* function. The exponent is the number of weeks, n.

$$f(n) = 2^n$$

Exponential Growth and Decay Functions: When a given quantity is increased or decreased over time by a certain percentage we can calculate anticipated results using one of these two formulas.

Growth: Original amount is increasing $A_f = A_0(1 + r)^t$

Decay: Original amount is decreasing $A_f = A_0(1 - r)^t$

A_f = Final Amount

A_0 = Original (or starting) amount

r = rate of increase or decrease for a specific time, in decimal form.

t = time (this must match the time units in the rate. If rate is per year, time must be in years. If rate is per month, or per week, time units must be months, or weeks.)

(Examples on the next page)

Examples Exponential Growth and Decay Functions

❶ Weeds are growing in Tony's front lawn at a rate of 5% per week. The lawn is 7500 square feet. If there are 40 square feet covered with weeds now, how many square feet, to the nearest integer, of the lawn will be covered with weeds after 7 weeks?

Hint: The amount of weeds at the end of 7 weeks is more than the original amount. This is an *exponential growth function.*

$A_f = ?$ $\quad A_0 = 40$ $\quad r = 5\% = 0.05$ $\quad t = 7$

$A_f = A_0(1 + r)^t$

$A_f = (40)(1 + 0.05)^7$

$A_f = 40(1.05)^7$

$A_f = 56.284$; $A_f \approx 56$ ***sq. feet***

❷ Kimberly's family has a pool. The pool is filled with water on June 1 in the morning. The pool holds 151,875 gallons of water. The water evaporates at an average rate of 0.5% per day in the month of June.

a) What is the volume, to the nearest gallon, of the pool in the morning on July 1? (June has 30 days.)

b) How many gallons of water would need to be added on July 1 to fill the pool again?

c) What is the total percent of pool water lost during that time?

Solution: This is an *exponential decay function* since the pool water is decreasing by 0.5% per day. 30 full days of evaporation is occurring. The original amount of water is 151,875 gallons. The other calculations can be made based on this result.

$A_f = ?$ $\quad A_0 = 151{,}875$ $\quad r = 0.5\% = 0.005$ *per day* $\quad t = 30$ *days*

Formula: $A_f = A_0(1 - r)^t$

$A_f = 151{,}875(1 - 0.005)^{30}$

$A_f = 130{,}670.8$ *gallons*

$A_f = 130{,}671$ gallons in a pool at the end of June.

$A_0 - A_f =$ *gallons evaporated*

$151{,}875 - 130{,}671 = 21{,}204$ *gallons needed to refill pool.*

$$\frac{\text{Loss by evaporation}}{\text{Full amount}} = \frac{\%\text{ lost}}{100\%}$$

$\dfrac{21{,}204}{151{,}875}$ g $100\% = 13.9\%$ *of water evaporated.*

Linear, Quadratic and Exponential Models

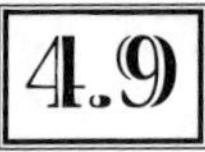

4.9

Creating a Function From a Table

Examples

❶ The accompanying table represents a savings account for the last year. Write the account balance as a function of the monthly savings.

Week	Balance
1	110
2	195
3	280
4	365
5	450
6	535
7	620
8	705
9	790
10	875
11	960
12	1045

Analysis: Determine what kind of function this table represents. For each increase of one week, w, the balance $f(w)$ increases by \$85. This is a *linear function*. The slope of the graph is 85.

The table does not give information about how much money was put in the savings account to start it. That would be week zero. Since week 1 shows a balance of \$110 and we know that is \$85 more than the week before, \$25 must have been the starting amount.

The account began with a balance of \$25 and each month, an additional \$85 was deposited into the account.

The domain is $x \geq 0$ since we can't use negative values for the weeks.

$$f(x) = 85x + 25$$

❷ A scientist is working on an experiment. The volume of the substance he is working with increases 5 times for every unit of heat that is added to the experiment. Write a function to represent his experiment.

h	$f(h)$
0	1
1	5
2	25
3	125
4	625
5	3125
6	15625

Analysis: In this table, for each increase of one unit of the independent variable, the dependent variable increases by a factor of 5. This is an *exponential function*.

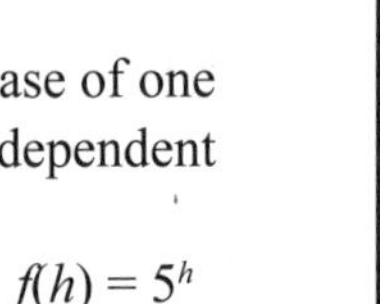

$$f(h) = 5^h$$

Question: If he adds 10 units of heat, what will the volume of the substance be?

$f(10) = 5^{10}$

$f(10) = 9{,}765{,}625$ *units of volume*

Observations From Graphs

In some circumstances it is necessary to make decisions about which type of function best suits a particular situation.

Example Danny is saving to buy something special. He has only \$1.00 to begin with. His Dad says he will give Danny \$100 every week to add to his account. His uncle offers to double the money in his account each week. Write a function and sketch a graph to demonstrate this situation. Determine which would be the better deal for Danny? Explain your answer.

Analysis: Write a function to represent each situation. Let x represent the number of weeks. Sketch the graphs of both functions and compare.

Dad: $f(x) = 1 + 100x$ Starts with \$1 and adds \$100 each week. This is a linear function.

Uncle: $g(x) = 2^x$ Initial amount of \$1 is doubled each week. This is an exponential function. x is an exponent in $g(x)$.

Graph both functions

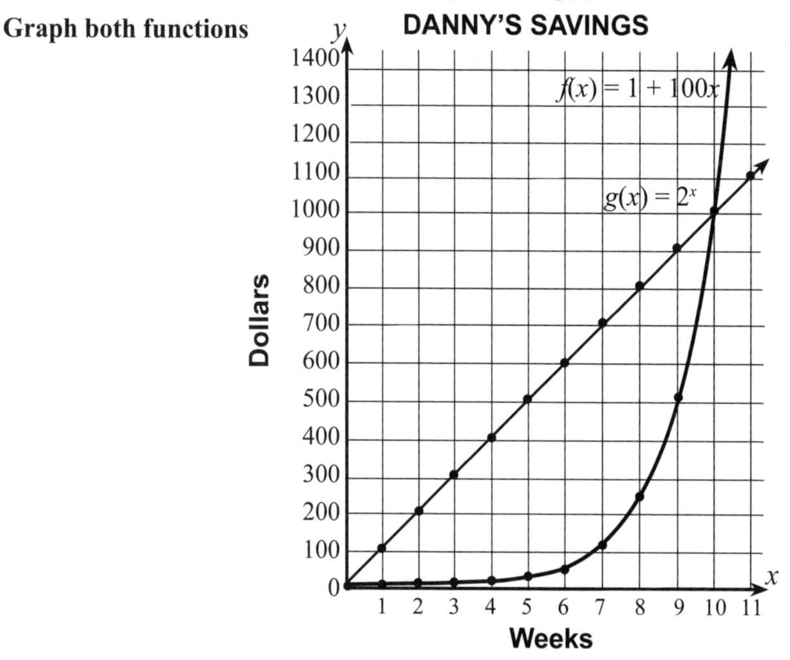

The two functions intersect at approximately ten weeks. After ten weeks the exponential function, $g(x)$, increases much more rapidly than the linear function, $f(x)$. If Danny is going to make his purchase in the first 10 weeks of his savings program, then he should take the offer his Dad made. If he is saving for longer than ten weeks then his uncle's plan is considerably better. At 11 weeks, $g(11) = \$2048$ while $f(11) = \$1101$.

4.10 COMPARING FUNCTIONS

Comparing the properties like maxima, minima, domain, range and other characteristics of different types of functions can be helpful in choosing which function is appropriate for use in a particular situation. As a result of the comparisons, determining which function will give the desired results can influence the choice made for solving a problem. Information about a function can be determined from graphs, tables, equations, or descriptions.

Examples

Figure 1

x	$g(x)$
–4	–16
–2	–4
0	0
2	–4
4	–16

❶ **Compare $f(x)$ which is given in (*a*) and $g(x)$ which is shown in the table. (*Figure 1*)**

a) $f(x) = x^2 + 5$

b) The points in the table (***Figure 1***) at the right represent several points in the function, $g(x)$.

Discussion:

a) $f(x)$ is a quadratic function with a minimum value of 5. Its axis of symmetry is the y-axis and its vertex is at (0, 5). Its maximum value is $+\infty$. The domain is the real numbers. The range is $y \geq 5$.

b) $g(x)$ is also a quadratic function. Its maximum value is 0 and its minimum value is $-\infty$. Its axis of symmetry is the y-axis. The domain is the real numbers. The range is $y \leq 0$.

❷ **Compare these two functions.**

a) Geraldine is saving money in a jar. She started out with $4 and is putting in $2.00 per week.

b) ***Figure 2b*** represents the amount of money Suzanne puts in her account on a weekly basis.

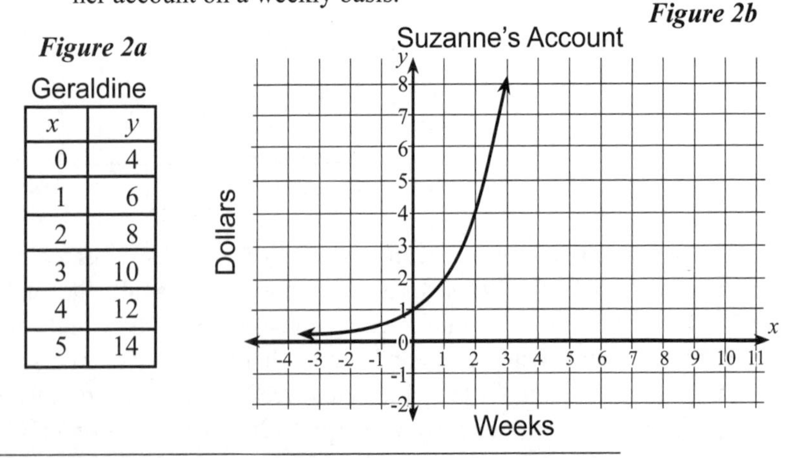
Figure 2a

Geraldine

x	y
0	4
1	6
2	8
3	10
4	12
5	14

Figure 2b

Discussion:

a) Geraldine starts with \$4 which is the minimum value of the function in part *a*. Each week her savings increase by \$2. Theoretically, the maximum value is $+\infty$. This is a linear function. The domain is $x \geq 0$ and the range is $y \geq 4$.

b) Suzanne could not begin at –2 weeks, so we must look at the value of the function in part *b* when $x = 0$ to find the minimum value. The minimum value is \$1 which indicates that is what she started with. The first week the value increased by 1, making the value \$2. The following week the value increased to \$4 which is double the previous value. Week 3 is consistent in that the value is double the week 2 value. This is an exponential function. Again, theoretically the maximum value of the function is $+\infty$. The domain is $x \geq 0$ and the range is $y \geq 1$.

- In comparing the two functions, it is clear that Suzanne will accumulate more money more quickly than Geraldine. For example, at the 5 week mark, Geraldine will have \$14 saved, while Suzanne will have \$32 dollars saved.

❸ **Compare these two functions.**

The coordinates in ***Figure 3a*** displays points in a function, 3*a*. The graph in ***Figure 3b*** represents function 3*b*. Compare the functions.

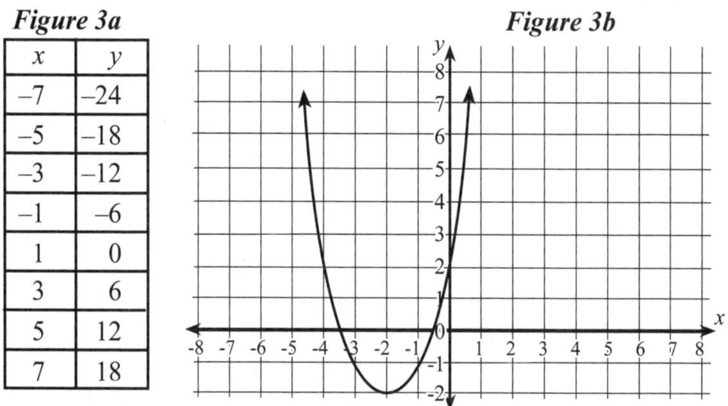

Figure 3a

x	y
–7	–24
–5	–18
–3	–12
–1	–6
1	0
3	6
5	12
7	18

Figure 3b

Discussion: Figure 3*a* represents a linear function. Figure 3*b* represents a quadratic function. In the interval [–3, 5] the function in 3*a* is increasing. In figure 3*b* in the interval [–3, 5] the function is decreasing when $x < -2$. and increasing when $x \geq -2$. The domain and range of function 3*a* are both all the real numbers. The domain of function 3*b* is all the real numbers and the range is $y \geq -2$.

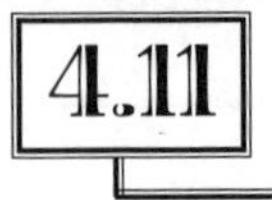

SEQUENCES

Sequence: A list of terms or elements in order. The terms are identified using positive integers as subscripts of a: $a_1\ a_2\ a_3 \ldots a_n$. The domain is the set of consecutive positive integers starting with 1. The terms in a sequence can form a pattern or they can be random.

Series: The sum of the terms of a sequence.

Term or Element: a number in a sequence.

Subscripts: Consecutive counting (natural) number subscripts, starting with 1, that are used to identify the location of the terms in the sequence. Each term is referenced by that subscript which is called the index of the term.

Example a_3 indicates the 3rd term in the sequence.

Domain: In a finite sequence the domain is a subset of the positive integers (counting or natural numbers). In an infinite sequence the domain is the set of all positive integers.

Range: The terms listed in the sequence form the range of the sequence.

Finite Sequence: Contains a specific number of terms that can be counted. The domain is a subset of the counting or natural numbers.

Examples

Sequence: 2, 5, 8, 11, 14, ... 72 ***Sequence:*** 5, 10, 20, 40, 80, ... 640
Index: $a_1, a_2, a_3, a_4, a_5, \ldots a_n$ ***Index:*** $a_1, a_2, a_3, a_4, a_5, \ldots a_n$

Infinite Sequence: Contains an unlimited number of terms that cannot be counted. The domain is the set of all natural numbers.

Examples

Sequence: 2, 5, 7, 12, 19, 24, 26, ... ***Sequence:*** 2, 4, 8, 16, 32, ...
Index: $a_1, a_2, a_3, a_4, a_5, a_6, a_7, \ldots$ ***Index:*** $a_1, a_2, a_3, a_4, a_5, \ldots$

Sequences with Patterns: An equation, also called a formula or a definition, can be used to work with a sequence. In some formulas specific numeric terms are not given. The formula depends on the index of the term desired – its location in the sequence. This type of formula is sometimes called an explicit formula or definition. A specific term can be found by substituting the number of the term for *n*. A recursive formula can be used when one or more of the terms of the sequence are given. In a recursive formula, a specific term is found by using terms located before it in the sequence.

SEQUENCES
Finding a specific term in a sequence

Recursive Definition or Formula: In a recursive definition or formula, the first term in a sequence is given and subsequent terms are defined by the terms before it. If a_n is the term we are looking for, a_{n-1} is the term before it. To find a specific term, terms prior to it must be found.

Example Find the first three terms in the sequence $a_n = 3a_{n-1} + 4$ if $a_1 = 5$. In this example, the first term is $a_1 = 5$. To find the second and third terms, $n = 2$, and $n = 3$ need to be substituted.

$a_1 = 5$
$a_2 = 3(a_1) + 4;\quad a_2 = 3(5) + 4;\quad a_2 = 19$
$a_3 = 3(a_2) + 4;\quad a_3 = 3(19) + 4;\quad a_3 = 61$

The three terms are 5, 19, and 61.

To *write a recursive definition* or formula when given several terms in the sequence, it is necessary to find an expression that is developed by comparing the terms and finding the process required to change each term to the subsequent term.

Example Write a recursive definition for this sequence. –2, 4, 16, 256, …
$\mathbf{a_1 = -2}$. Since $4 = (-2)^2$, and $16 = 4^2$, and using the last term given to us, $256 = 16^2$, a recursive definition for this sequence could be $a_n = (a_{n-1})^2$

Explicit Formula: If specific terms are not given, a formula, sometimes called an explicit formula, is given. It can be used by substituting the number of the term desired into the formula for "n". Simplify as usual.

Examples

❶ What is the seventh term in the sequence $a_n = 2n - 4$
Since we want the seventh term, $n = 7$.
Substitute 7 in place of n in the equation. $a_7 = 2(7) - 4$
The seventh term in this sequence is 10. $a_7 = 10$

❷ What is the fifth term of the sequence $a_n = 3^n$?
Substitute 5 for n. $a_5 = 3^5$
The fifth term in this sequence is 243. $a_5 = 243$

❸ What are the first three terms in the sequence $a_n = n^2 + 1$?
Three calculations are needed: $n = 1$, $n = 2$, and $n = 3$.

$a_1 = 1^2 + 1 = 2$
$a_2 = 2^2 + 1 = 5$ The first three terms are: 2, 5, 10
$a_3 = 3^2 + 1 = 10$

Note: Terms should be in simplified whenever possible.

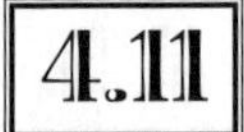

ARITHMETIC SEQUENCE

Each term in the sequence has a common difference, ***d***, with the term preceding it. The first term is labeled a_1. The formula for finding *specific terms of an arithmetic sequence* is $\mathbf{a_n = a_1 + (n - 1)d}$, where a_n is the term desired, a_1 is the first term in the sequence, n is the location in the sequence of the term desired, and d is the common difference.

To find *d*, COMMON DIFFERENCE: $a_2 - a_1$, $a_3 - a_2$, etc.

If given the first term and the value of d, the formula can be used to find other terms in the sequence.

Examples

❶ Find the seventh term of an arithmetic sequence if $a_1 = 5$ and $d = 2$.

$$a_n = a_1 + (n - 1)d$$
$$a_7 = 5 + (7 - 1)(2)$$
$$a_7 = 5 + 12$$
$$a_7 = 17$$

❷ In an arithmetic series $a_1 = 5$. Find a_{10} if $a_6 = 17$ and $a_7 = 19$.

Find d: $a_7 - a_6 = 19 - 17$; $d = 2$

Use formula: $a_n = a_1 + (n - 1)d$

$$a_{10} = 5 + (9)(2)$$
$$a_{10} = 23$$

Recursive Formula: Terms in the sequence are given to establish a pattern. The *general recursive formula* for an arithmetic sequence is $\mathbf{a_n = a_{n-1} + d}$ but we have to find d.

Example Write the formula for this sequence. {1, 4, 7, 10, 13 …}

There is a common difference, d, of 3 between each pair of consecutive terms in the sequence. Each term in the sequence is found by adding 3 to the previous term. Since the terms were given, a *recursive* formula can be developed. $d_1 = 3$. $a_n = a_{n-1} + 3$.

Find the sixth term of this sequence: $a_6 = a_5 + 3$; $a_6 = 13 + 3 = 16$.

The sixth term of this sequence is 16.

To find the twentieth term of this sequence we would need the nineteenth term to use this formula. It would make more sense to use the explicit formula, $a_n = a_1 + (n - 1)d$

$a_{20} = a_1 + (20 - 1)(d)$; $a_{20} = 1 + 19(3)$; $a_{20} = 58$

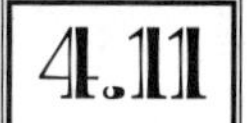

GEOMETRIC SEQUENCE

The consecutive terms are developed by multiplying each term in the sequence by a common ratio, r, to obtain the next consecutive term. The terms in the sequence have a common ratio, $r = \frac{a_2}{a_1}$. (Some texts refer to a geometric sequence as a geometric progression).

To find r, the COMMON RATIO: Divide a term by the term before it. $r = \frac{a_2}{a_1}$; $r = \frac{a_4}{a_3}$… Any two consecutive terms in a geometric sequence will have the same common ratio.

Examples

❶ 3, 6, 12, 24 ... (a_1 a_2 a_3 a_4) Each pair of terms has a ratio of 2. $\frac{6}{3} = 2,\ \frac{12}{6} = 2,\ \frac{24}{12} = 2.$

Each term in this sequence is found by multiplying the previous term by 2.

❷ $\frac{27}{8}, \frac{9}{4}, \frac{3}{2} \ldots$ (a_1 a_2 a_3) $r = \frac{\frac{9}{4}}{\frac{27}{8}} = \frac{2}{3},\quad r = \frac{\frac{3}{2}}{\frac{9}{4}} = \frac{2}{3}$ ***Common Ratio r*** $= \frac{2}{3}$

Each term in this sequence is multiplied by 2/3 to get the next term.

Finding a Specific Term of a Geometric Sequence: Use the formula $a_n = a_1 r^{n-1}$ where n is the index of the term desired, r is the common ratio of the sequence and a_1 is the first term of the sequence. Determine the value of r first, then substitute and simplify.

Using the sequence in example 1 above, 3, 6, 12, 24, find the 12th term

$n = 12, r = 2, a_1 = 3$

$$a_{12} = a_1 r^{n-1}$$
$$a_{12} = (3)(2^{(12-1)})$$
$$a_{12} = 3(2048)$$
$$a_{12} = 6144$$

Find the seventh term in this sequence $\frac{1}{2}, \frac{1}{4}, \frac{1}{8} \ldots$ ***First find r : r*** $= \frac{\frac{1}{4}}{\frac{1}{2}} = \frac{1}{2}$

$a_7 = \frac{1}{2} \cdot \left(\frac{1}{2}\right)^6;\quad a_7 = \left(\frac{1}{2}\right) \cdot \left(\frac{1}{64}\right);\quad a_7 = \frac{1}{128}$

The Recursive Formula for a geometric sequence is $\boldsymbol{a_n = (a_{n-1})r}$ when terms are given.

Example What is the fifth term of the sequence 5, 10, 20, 40,…

$r = \frac{20}{10} = 2,\ \frac{40}{20} = 2;\ r = 2,\ a_{n-1} = 40$ 40 *is the fourth term*

$a_5 = (40)(2) = 80$

Building Functions

Descriptive Statistics

- Summarize, represent, and interpret data on a single count or measurement variable.
- Summarize, represent, and interpret data on two categorical and quantitative variables.
- Interpret linear models.

5.1

Statistics is the mathematics of collecting, organizing, summarizing, and analyzing data. The data are often displayed using graphs and tables. Data collected involve *individuals* which are the objects described in the data. They can be people, animals, test scores, measurements, or other items. A *variable* is the term used to describe the characteristic of an individual. This may have different values for different individuals. Example: If the data represent the weight of 30 dogs, each dog is an individual and the variable is the weight of the dog. If the data describe the favorite movie of each of 30 students, each student is an individual and the name of the movie is the variable.

TYPES OF VARIABLES AND DATA

Categorical Variable: Allows the identification of the group or category in which the individual is placed.

Quantitative Variable: Results are numerical which allows arithmetic to be performed on them.

Categorical Data is non-numerical. The data values are identified by type. It is also called qualitative data.

Example Eye color, kind of pet owned, or the name of a favorite TV show or movie.

Quantitative Data is numerical. The data values (or items) are measurements or counts and have meaning as numbers.

Example Grades on a test, hours watching TV, or heights of students.

Univariate: Measurements are made on only one variable per observation.

Example
- Quantitative: Ages of the students in a club.
- Categorical: Kind of car owned.

Bivariate: Measurements are made on two variables per observation.

Example
- Quantitative: Grade level and age of the students in a school.
- Categorical: Gender and favorite T.V. shows.

Biased: A data set that is obtained that is likely to be influenced by something – giving a "slant" to the results.

Example
- Quantitative: To determine the average age of high school students by asking only tenth graders how old they are.
- Categorical: Standing outside Yankee Stadium and asking people coming out of the stadium to name their favorite baseball team. Most would say … Yankees!

Unbiased: A data set that is obtained which has no connection to anything that would influence the results.

Example
- Quantitative: Asking people coming out of a stadium how many pets they have.
- Categorical: Asking people leaving a large grocery store what their favorite flavor of soda is.

CHARACTERISTICS OF CATEGORICAL AND QUANTITATIVE DATA

Single (ungrouped): Small samples or collections of data that can be treated as individual items. Statistical information such as mean, median, or mode, can be obtained by listing the data values individually.

Example
- Quantitative: Test scores of 42, 65, 65, 70, 72, 75, 80.
- Categorical: In a classroom of students what pet(s) each student has.

Grouped: When large numbers of values are included, the data are often treated in groups called intervals. The range of the data is divided into equal intervals and each item of data is recorded in the appropriate interval. The statistical information is located in the intervals.

Example
- Quantitative: Heights of 200 students in a school. Their heights range, in inches, from 56″ to 75″. The intervals could be 56-60, 61-65, 66-70, 71-75. Each of the 200 students has a height within one of these intervals.
- Categorical: Within certain age groups in a city, the highest level of education achieved – high school, trade school, 2 year college, 4 year college, post-graduate, doctorate.

ORGANIZING QUANTITATIVE DATA

RECORDING AND ANALYZING QUANTITATIVE DATA

Tally Sheet: A chart that is used to count the number of items with the same value (for single data items) or the items whose value falls in each interval (grouped data.) The tally sheet is marked off with small lines and every 5th tally is crossed through the previous 4 so a unit of 5 is formed. (See page 154)

Frequency Table: A completed tally sheet which shows the intervals and the frequencies of each data item. (See page 154)

Cumulative Frequency Table: Shows the sum of the frequencies at or below each of the intervals in the frequency table. The accumulation usually begins with the interval nearest zero. (See page 154)

QUANTITATIVE DATA CALCULATIONS

Use the test scores 42, 65, 65, 70, 72, 75, 80

Range: Highest value in the data set minus the lowest value.

Example $80 - 42 = 38$. Range = 38.

Mean: Average. The sum of the data values divided by the number of items in the data.

Example $42 + 65 + 65 + 70 + 72 + 75 + 80 = 469$.
Divide by 7. Mean = 67.

Mode: Value that appears most often. 65 appears twice, the mode = 65. If the data has 2 modes, it is called ***bimodal***. Sometimes there is no mode and the data are described as having ***no mode***.

Median: The middle data value when the data are listed in order, ascending or descending. Find the number of items, n, and divide by 2 to find the *position* of the middle piece of data. If n is even, find the numerical average of the values of the two middle items. In our example, $n = 7$. The middle item is the 4th item and its value is 70. The median = 70.

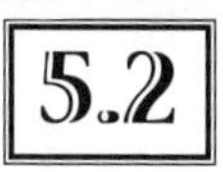

Percentiles and Quartiles: These measures show how a particular item of data compares to the other items in the data.

Example If a person's height is in the 90th percentile for his age, that means at approximately 90% of the people his age who were measured were the same height or shorter than he is.

Quartiles: Quantitative data can be divided into sections for various analyzing purposes. When using grouped data, the interval that contains the specific quartile is used. (See also page 155)

Middle Quartile or 50th Percentile: The value of the median.

Lower or 1st Quartile or the 25th Percentile: The median of the values below the middle quartile.

Upper or 3rd Quartile or the 75th Percentile: The median of the values above the middle quartile.

Interquartile Range (IQR): The difference between the value of the upper quartile and the lower quartile.

Examples

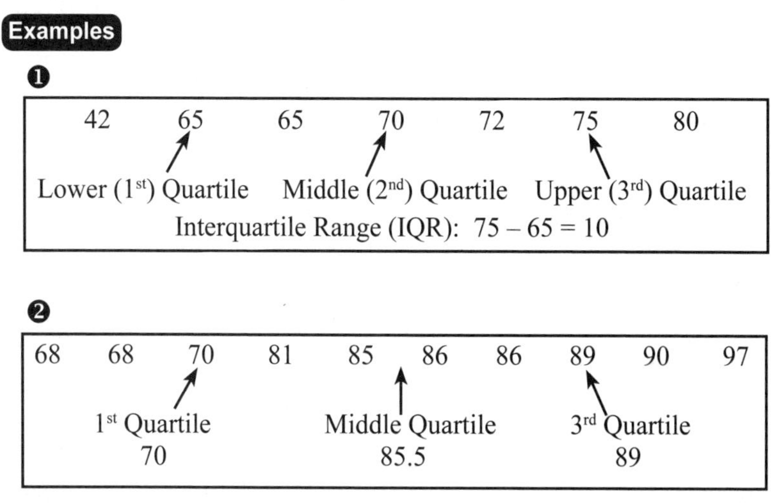

When an even number of data items are used, the median or middle quartile is the average of the two middle values.

(See also page 158 for Outlier(s).)

Interpreting Categorical & Quantitative Data

DISPLAYING UNIVARIATE DATA (ONE VARIABLE)

Depending on the distribution (the way the data occur) various types of graphs are used to display the data.

Bias in Graphing: It is important when representing data on a graph to be sure the graph itself is an accurate presentation of the data. Two ways a graph can be biased are if the scales are not appropriate, or if there is a break in the presentation of the data.

Histogram: Data that are grouped in intervals are often displayed using a histogram. It is similar to a bar graph, but the "bars" are adjacent to each other and touching. The horizontal axis is labeled with the intervals; the vertical axis is the frequency, or the cumulative frequency if that is the type of histogram being used. Remember, the intervals must be of equal length, so the horizontal scale must be divided into equal segments so the "bars" are all equal in width. A histogram is not usually used for ungrouped sets of data.
(See dot plots on page 157 and scatter plots on page 163.)

Example Make a frequency table, cumulative frequency table, and appropriate histograms for the following data. Using the tables created, determine the intervals of the mean, median, and mode.

Student grades: 82, 84, 84, 75, 78, 68, 68, 98, 87, 84,
86, 92, 68, 87, 89, 75, 66, 78, 89, 90.

Mean: Add the numbers listed, then divide by the number of items (20): $1628 \div 20 = 81.4$. 81.4 is in the interval 81-85

Median: List the numbers in order, smallest to largest. There are 20 numbers so the median is the average of the tenth and eleventh items. Since both are 84, the median is also 84. The interval is 81-85.

Mode: The interval containing the mode is 86-90.

FREQUENCY TABLE			CUMULATIVE FREQUENCY TABLE	
Interval	**Tally**	**Frequency**	**Interval**	**Cumulative Frequency**
96 – 100	I	1	66 – 100	(19 + 1) = 20
91 – 95	I	1	66 – 95	(18 + 1) = 19
86 – 90	~~IIII~~ II	7	66 – 90	(11 + 3) = 18
81 – 85	III	3	66 – 85	(8 + 7) = 11
76 – 80	II	2	66 – 80	(6 + 2) = 8
71 – 75	II	2	66 – 75	(4 + 2) = 6
66 – 70	IIII	4	66 – 70	4

FREQUENCY HISTOGRAM

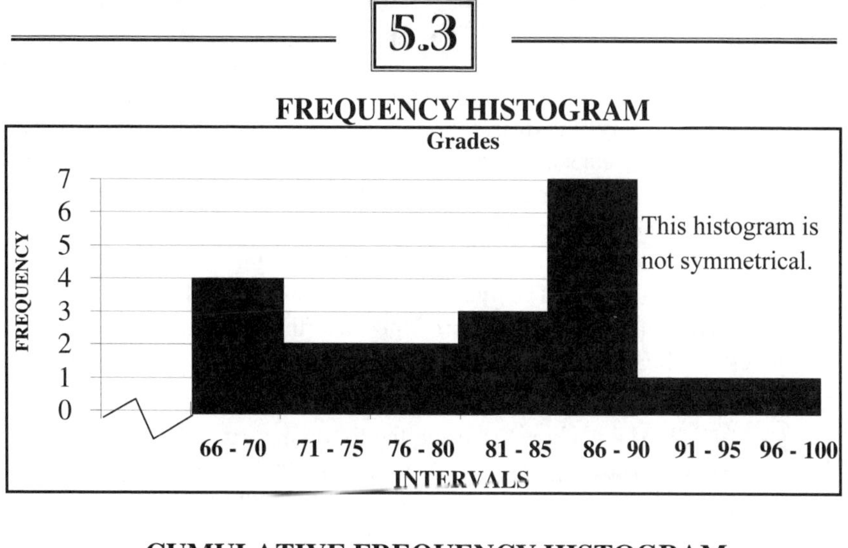

CUMULATIVE FREQUENCY HISTOGRAM

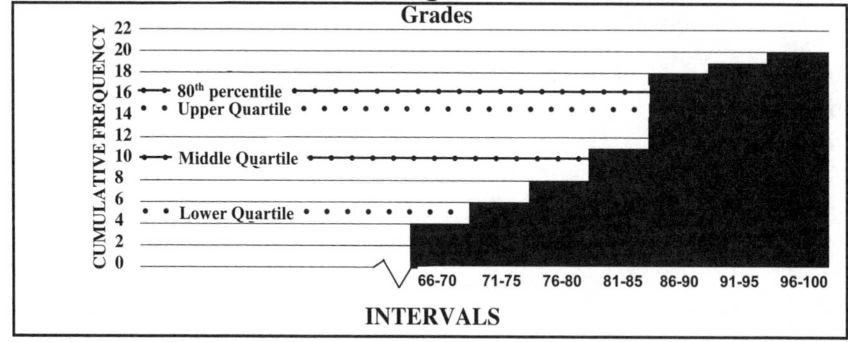

QUARTILES AND PERCENTILES

Lower: (.25) (20) = 5	The 5th item is in the 71-75 interval.
Middle: (.50) (20) = 10	The 10th item is in 81-85 interval.
Upper: (.75) (20) = 15	The 15th item is in the 86-90 interval.
80th percentile: (.80) (20) = 16	The 80th percentile is the 16th item counting from the bottom of the range. it is located in the interval 86 –90.

Box Plot: This graph is sometimes called a *box and whisker plot*. It is a graph that displays the distribution in four portions using 5 numbers that summarize the data set as the boundaries of the sections. The ***5 summary numbers*** include the smallest and the largest observed values, the median, and the upper or third quartile and lower or first quartile. The left and right ends of the center box represent the first and third (lower and upper) quartiles respectively. The line inside the box identifies the median. Lines extending out from both sides of the box (the whiskers!) indicate the lowest and highest values observed. The box plot does not show individual data values.

Example Make a box plot using the following data:

Student grades: 82, 84, 84, 75, 78, 68, 68, 98, 87, 84, 86, 92, 68, 87, 89, 75, 66, 78, 89, 90.

First put them in order from smallest to largest for ease in working with this graph and the dot plot which follows.

66 ← Lowest Value = 66
68
68
68
75
75 ← Lower or 1st Quartile = 75 (The median of the lower half of the data)
78
78
82
84
84 ← Median = 84. Since this is an even number of values, the median is halfway between the middle two values. Since both of these are 84 in this example, the median is 84.
84
86
87
87
89 ← Upper or 3rd Quartile = 88. The average of the 5th and 6th terms in the upper half of the data.
89
90
92
98 ← Highest value = 98

Student Grades

65 70 75 80 85 90 95 100

Dot Plot: A number line is created for the horizontal axis of this graph and each item of data is noted above the appropriate number. The notation can be in dots, **x**'s, or other non-numerical marks. While the box plot shows the distribution and the 5 summary points of the data, the dot plot allows us to see the individual values and how many values there are in the data set.

Example Using the same data as we did for the box plot on the previous page, make a dot plot. Arrange the grades in order to make it easier to work with.

66, 68, 68, 68, 75, 75, 78, 78, 82, 84, 84, 84, 86, 87, 87, 89, 89, 90, 92, 98

Steps

1) determine the highest and lowest values and choose a convenient labeling for the horizontal axis. In this problem the range is 98 – 66 = 32. It seems that labeling by 5's may be good.
2) Draw AND label the horizontal axis like a number line.
3) Record each data value with a dot (or an x) above the number line.

STUDENT GRADES

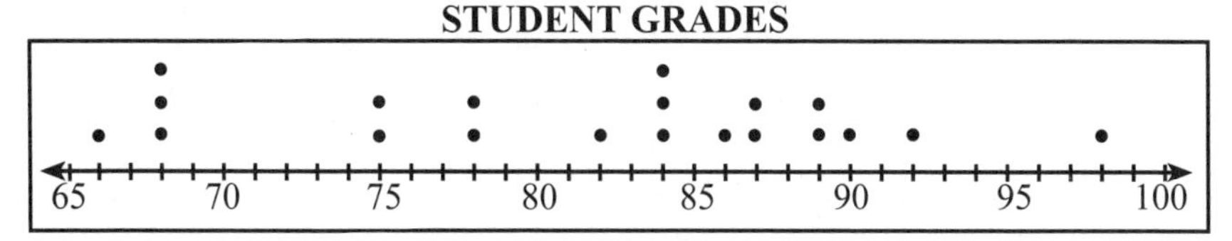

DISTRIBUTIONS AND COMPARISONS

Shapes of Distributions: When data are plotted the overall general shape of the distribution can sometimes take a form that can be described in a few words. (See sketches on the next page.)

Center: A value that divides the data so that about half of the points are smaller than the center and about half are larger. Two common measures of the center are the mean and the median.

Range: The difference between the highest and lowest data values.

Interquartile Range: The difference between the lower quartile and the upper quartile. Often abbreviated as IQR.

Spread: The spread of the data can best describe how the data are dispersed; whether they are clustered or not clustered together. Consider both the range and the interquartile range when describing the compactness of the data, as well as the visualization of the data on a graph. The type of graph used is often a box plot, dot plot, or histogram. (See also page 155 for histograms)

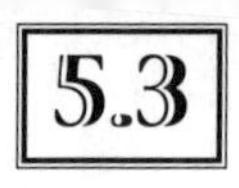

Symmetrical: If the graph is folded in half vertically at the center and the two sides match, the distribution is called symmetrical. The amount of data on one side of the central line relatively matches that on the opposite side. In a symmetrical distribution, the mean and the median are close together or may even be the same value.

Skewed: If the data have an item (or several items) of data that is located significantly distant from the main grouping of data, the general shape of the data may appear to have a "tail". If the tail goes toward the right, the data are skewed right. If the tail goes toward the left, the data are skewed left; the data are not symmetric. The low or high value (tail) can have a strong effect on the mean which then is not usually an accurate measure of the center of the data.

Normal: If a curve sketched along the highest data values on the graph has the appearance of the familiar bell shaped curve, it is a normal distribution.

Outlier: Data point(s) far outside the main grouping of the data are called outliers. Outliers can strongly affect the mean. In some cases an outlier is an error in the recording of the data, in others it is a correct data item. Outliers are sometimes included and sometimes excluded from the statistical calculations, depending on the type of data recorded and the circumstances of the data set. Where outliers exist and are included, they can have a strong influence on the mean as a result the mean is not a good measure of central tendency for a data set with outliers.

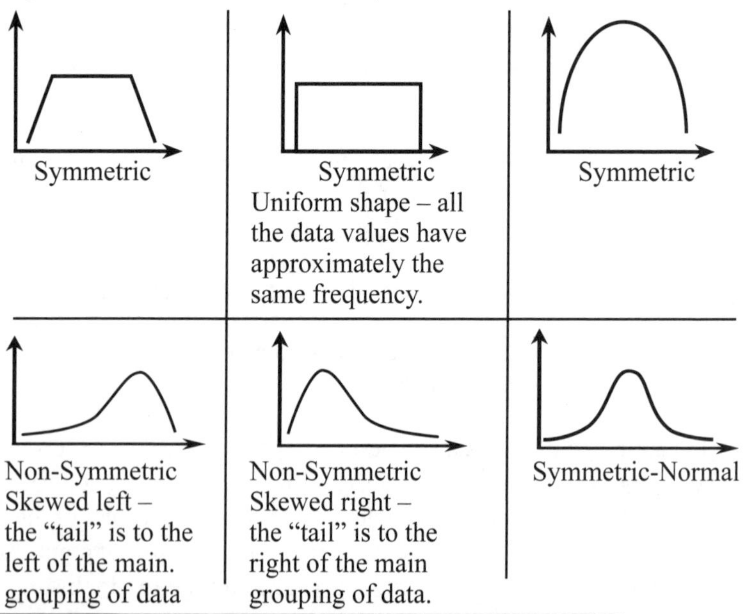

Examples The following three examples demonstrate data shapes.

❶ **Data Set 1:** Pre-Assessment test in Algebra class of 15 students. Grades are 20, 25, 26, 26, 30, 30, 30, 30, 32, 32, 32, 35, 40, 67, 70

Lowest value: 20
Lower/1st Quartile: 26
Median: 30
Upper/3rd Quartile: 35
Outlier: none*

Highest value: 70
Range: 50
Mode: 30
Mean: 35

* Inclusion or exclusion of a *possible* outlier(s) depends on the circumstances.

- Distribution is **non-symmetric**.
- Shape of data is **skewed right**. A "tail" appears to the right of the main data group.
- Most appropriate measure of central tendency is the **median**, because the high data points of 67 and 70 affect the mean.

PRE-ASSESSMENT GRADES

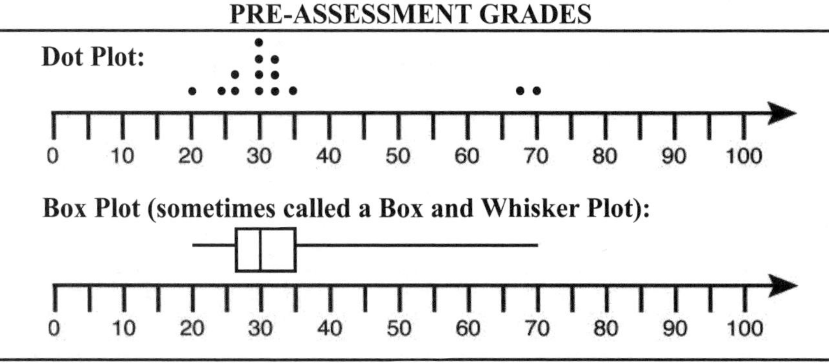

Questions for Data Set 1:

1. What are the values of the lower and upper quartile?
   Answer: 26 is the lower quartile value and 35 is the upper quartile value.
2. What is the range of the data?
   Answer: 50. The range can also be called the degree of dispersion of the data.
3. What is the Interquartile Range (IQR) of the data?
   Answer: The IQR is the difference between the upper and lower quartiles. 35 – 26 = 9
4. When thinking about describing the graph, which is the best measure of spread? Interquartile range or range?
   Answer: The range of the data is 50 and the interquartile range is 9. Most of the data are clustered about the median as shown by the IQR, but there are some data values (67 and 70) that indicate a much larger spread. The range gives a more accurate idea of the spread of the data because they are skewed.

❷ **Data Set 2:** Midterm test in Algebra class of 15 students.
Grades are 67, 70, 72, 72, 72, 75, 75, 75, 75, 75, 78, 78, 78, 80, 83

Lowest value: 67
Lower/1st Quartile: 72
Median: 75
Upper/3rd Quartile: 78

Highest value: 83
Range: 16
Mode: 75
Mean: 75

- Distribution is **symmetric**.
- Shape of data is **close to normal**. (See page 158)
- An appropriate center can be either the **mean or the median** since the data are symmetric.

MIDTERM TEST GRADES

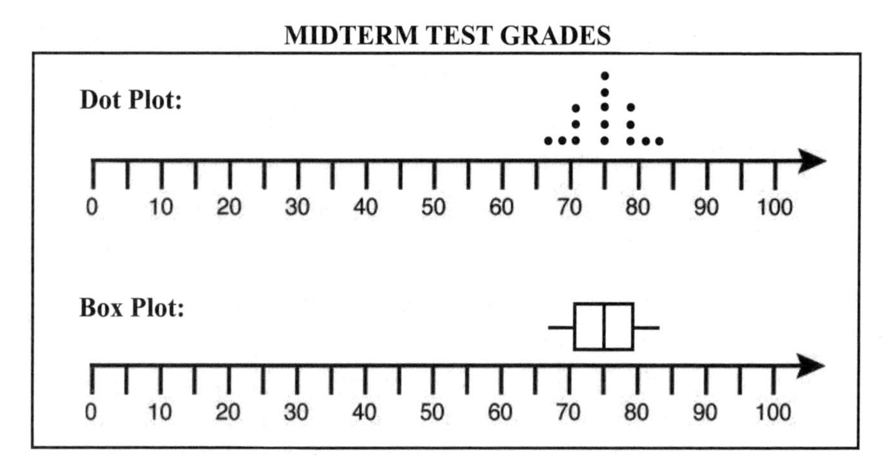

Questions for Data Set 2:

1. What are the values of the lower and upper quartile?
   Answer: 72 is the lower quartile value and 78 is the upper quartile value.

2. What is a degree of dispersion (range) of the data?
   Answer: 83 – 67 = 16 points.

3. What is the Interquartile Range (IQR) of the data?
   Answer: The IQR is the difference between the upper and lower quartiles. 78 – 72 = 6

4. When thinking about describing the graph, which is the best measure of spread? Interquartile range or range?
   Answer: The data values are very close to the mean or center of the data. The IQR appears to be the better measure of spread.

❸ **Data Set 3:** Final Exam in Algebra class of 15 students.
Grades are 65, 70, 78, 80, 83, 85, 85, 85, 85, 85, 87, 87, 87, 90, 90

Lowest value: 65
Lower/1st Quartile: 80
Median: 85
Upper/3rd Quartile: 87

Highest value: 90
Range: 25
Mode: 85
Mean: 82.8

Spread: 87 – 80 = 7
Outlier: none

- Distribution is **non-symmetric**.
- Shape of data is **skewed left**.
- Most appropriate center (or measure of variability) is the **median** because the data are skewed.

FINAL EXAM GRADES

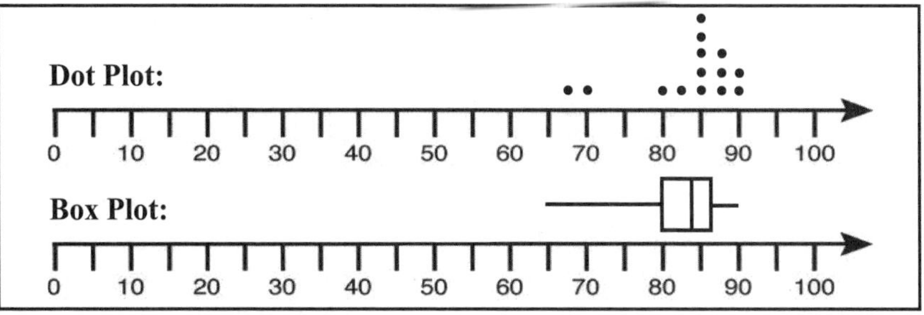

Questions for graph #3:

1. What is the value of the lower and upper quartile?
   Answer: 80 is the lower quartile value and 87 is the upper quartile value.
2. What is the degree of dispersion of the data?
   Answer: 90 – 65 = 25 points. The degree of dispersion can also be called the range of the data.
3. What is the Interquartile Range (IQR) of the data?
   Answer: The IQR is the difference between the upper and lower quartiles. 87 – 80 = 7
4. When thinking about describing the graph, which is the best measure of spread? Interquartile range or range?
   Answer: The range of the data is 25 and the interquartile range is 7. The range gives a more accurate idea of the spread of the data. Most of the data are very close to the middle, the median.
5. Why does the skewness of the data affect the choice of mean or median for the best measure of central tendency (center). Use specifics from example 3 to explain.
   Answer: The two considerably lower scores, 67 and 70, influence the numerical value of the mean strongly. The median is the data value in the center position, therefore its numerical value is not unduly influenced by the high scores.

In order to answer questions comparing box plots, it is sometimes helpful to graph them together.

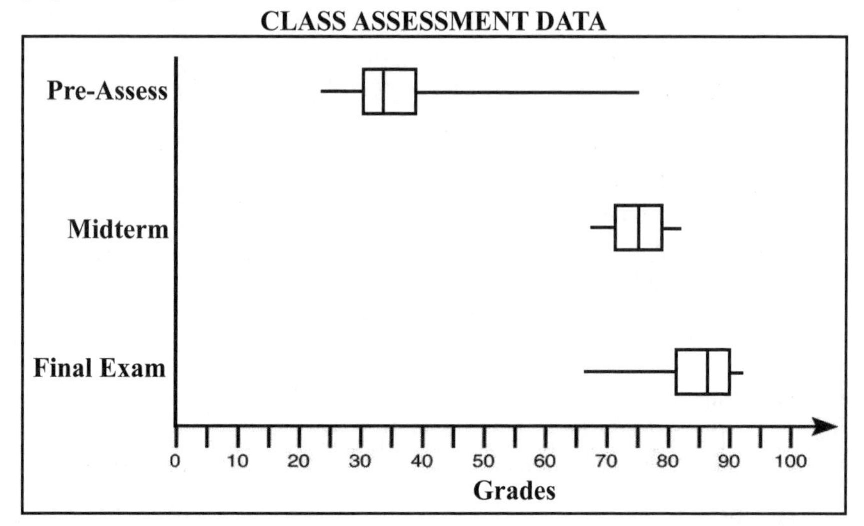

For comparing all three graphs:

1. Consider the shape of the data distribution to compare the best measure of center (or central tendency) for each graph. Which is the best measure of center?
   Answer: The mean is the best measure of center for the midterm scores. The data are more symmetric. The median is the best measure of center for the pre-assessment and for the final exam as the data are non-symmetric.
2. Which graph has the highest 50th percentile?
   Answer: The 50th percentile is also the median of the data. The highest median is the 3rd set of data with a median of 85.
3. Which set of data is the most greatly dispersed as measured by its interquartile range?
   Answer: Look for the graph with the longest "Box" to compare the interquartile ranges.
   Data set #1–IQR = 9; Data set #2–IQR = 7; Data set #3–IQR = 7 so data set #1 has the largest interquartile range and, therefore, data set #1 is the most dispersed.
4. Which set of data has the largest dispersion as measured by its range?
   Answer: Look for the graph with the longest "Box and Whisker Plot" length to compare ranges.
   Data set #1 Range = 50, Data set #2 Range = 16, Data set #3 Range = 25 so data set #1 has the largest range.

BIVARIATE DATA

DISPLAYING BIVARIATE DATA (2 VARIABLES)

The display of bivariate data depends on the type of data, quantitative or categorical, that is contained in the data set. Quantitative is discussed below. Categorical bivariate data can be displayed using a two-way frequency table. (See page 169)

Quantitative Data with two variables can be displayed using a **scatter plot**. In this numerical data, the values for the two variables are paired, much like (x, y) in coordinate graphing. Each pair of values becomes a point that is recorded on a grid using the horizontal axis for the independent value, x, and the vertical axis for the dependent variable, y. The graph displays the relationship between the independent and dependent variables. The plotted points often suggest a pattern (a line or a curve) which can be described using a function. Although producing a scatter plot and doing associated work with it can be done without a graphing calculator, it is recommended that one be used.

Line (or Curve) of Best Fit: It is a sketch through the points on the scatter plot such that the data points are distributed as equally as possible on both sides of the line or curve. A function (or equation) can be written to describe the line of best fit. This equation is also called a regression equation. In the example below, "Feeding the Birds", it seems that a line could be sketched through the data so the data points are equally distributed on each side of thc line. The line can be defined using a linear function. (Sometimes a curve is needed. See page 167.)

Regression Equation: A regression equation is a function that describes the data. It is also called the line (or curve) of best fit. Examine the scatter plot to see if the data appear to cluster around a line or curve. A regression equation can be written to describe the line or curve either manually or using a graphing calculator. The regression equation can also be used to predict the location of a data point not included in the observation.

Example Several neighbors were comparing the number of bird feeders that they have in their yards in the winter with the average amount of bird seed they used in a week.

Steps

1) Create a scatter plot to demonstrate the relationship between the number of bird feeders and the pounds of bird seed used in an average week.

2) Sketch a line or curve of best fit. Determine a function to define it.

3) Describe the relationship of the independent and dependent variables.

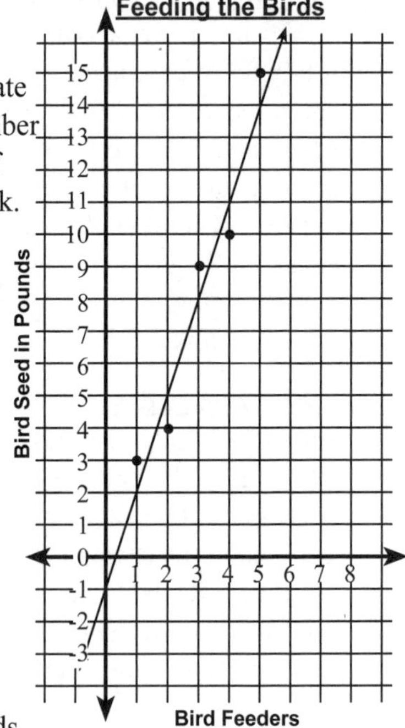

Bird Feeders	Bird Seed (pounds)
1	3
2	4
3	9
4	10
5	15

Solution:

a) The scatter plot for Feeding the Birds is sketched. Appropriate labels are included.

b) The data appear to cluster along a line. Sketch a line as accurately as possible between the data - some points maybe on it, some will not. After the line is sketched, it is necessary to define a function to describe this line.

c) When examining the data, as the independent variable (x) increases the dependent variable (y) also increases. This shows an upward or positive trend of the data.

Note: The line of best fit can be sketched by hand although the graphing calculators will make this process easier. Without the calculator the work needs to be done manually. Write an *approximate* equation of the line of best fit in the form $y = mx + b$.

FINDING THE APPROXIMATE REGRESSION EQUATION

Method Without A Calculator:

Steps

1) Choose two plotted points on the sketch that are close to or on the line of best fit. Use the slope formula, $m = \dfrac{y_2 - y_1}{x_2 - x_1}$, to find the slope of the line.

2) Then use the point-slope formula, $y - y_1 = m(x - x_1)$, to complete the equation. Write the equation $y = mx + b$ form.

Example The points closest to the line of best fit appear to be (1, 3) and (3, 9).

1) $m = \dfrac{9-3}{3-1} = \dfrac{6}{2} = 3$ *slope*

2) $y - 3 = 3(x - 1)$

$y - 3 = 3x - 3$

$y = 3x$ *approximate equation*

Method With A Calculator: The graphing calculator will provide a more accurate equation describing the line of best fit. When calculated using the appropriate features on thc graphing calculator, the equation of the line of best fit is $y = 3x - 0.8$. This is a *more accurate* estimation than the one done manually. The calculator equation will be used for the remaining work on this problem.

Calculator Screens: Shown here are the screens involved in doing this work using the TI-84 Plus Calculator.

Steps

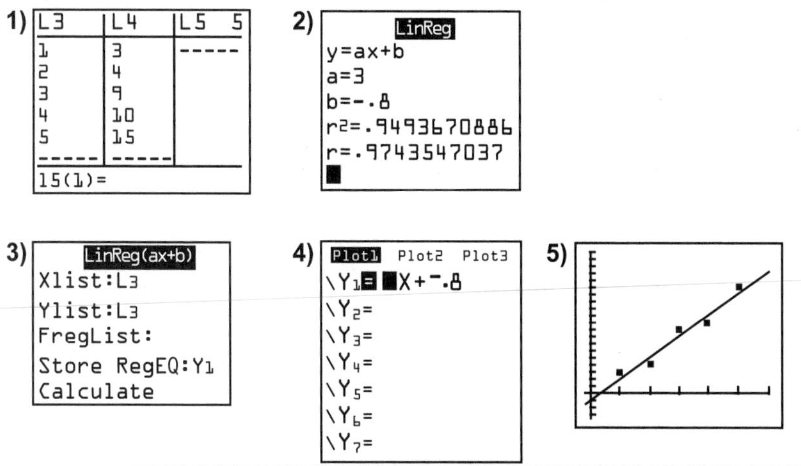

Prediction: The regression function can be used to *predict other points* within or beyond the observed data.

- **Interpolation:** Prediction within the observed data. If it were possible to have $2^1/_2$ bird feeders, substitute 2.5 for in the regression function. $f(x) = 3x - 0.8$. The amount of seed predicted would be about 6.7 lbs.
- **Extrapolation:** Prediction of a point beyond the observed data. To see how many pounds of bird seed should be needed for 50 bird feeders, substitute 50 for x in the equation $f(x) = 3x - 0.8$. The bird seed needed would be approximately 149.2 pounds.

Slope and y-intercept of the linear regression line with regard to the data must be considered. The regression equation calculated using the calculator is $y = 3x - 0.8$. ***The slope is 3 and the y-intercept is –0.8.*** Considering the type of data we collected, we know that a y-intercept of –0.8 is not possible. We know that we cannot have a negative value of a quantity of birdseed. (Remember this is an approximate equation.) The slope of 3 indicates that as the number of bird feeders increases by one, the pounds of bird seed needed increases by 3. Since the person who only has one feeder uses 3 pounds of seeds, it is reasonable that for each additional feeder 3 more pounds of seed are needed.

Correlation: It is a number that indicates how closely the data are represented by the line or curve of best fit. The number is called the correlation coefficient and is represented using r. The correlation coefficient, r, is part of the work done within the calculator to find the correct regression equation for the function. It is always between –1 and +1. The closer r is to –1 or to +1, the stronger the correlation. A weak correction is shown by an r value close to zero, either positive or negative.

Example In the bird feeder example. What is the correlation coefficient between the data and the line of best fit as determined by the calculator?

Solution: Use the statistics functions in your calculator to find the value of r.

The correlation coefficient $r = 0.974354037$. This number is quite close to +1 and shows a strong correlation between the data and the regression line. It shows a good fit for the line of best fit!

Positive or Negative Value of *r*: *The positive or negative value of **r** refers to the general trend of the data.*

- **Positive (+) correlation** means the data have an upward tendency, left to right, when graphed. A graph with a positive correlation will have a regression equation with a positive slope or a positive rate of change (slope between any two points on the graph). The bird feeder example data has a positive trend – as the independent variable increases, the dependent variable also increases.

Positive Correlation

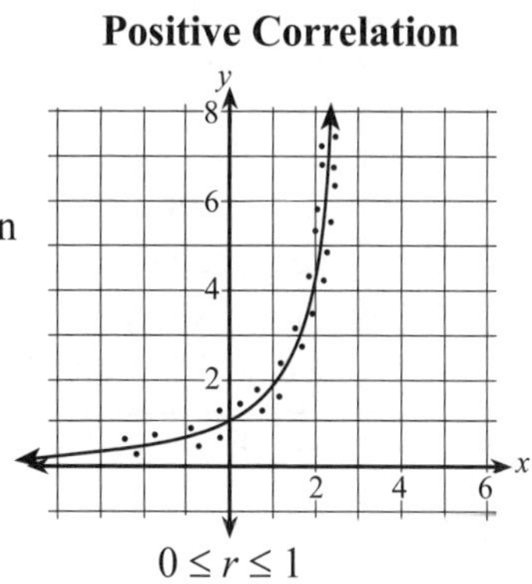

$0 \le r \le 1$

- **Negative (–) correlation** means the data have a downward tendency when read left to right on the graphed. A graph with a negative correlation will have a regression equation with a negative slope or a negative rate of change (slope between any two points on the graph). As the independent variable increases, the dependent variable decreases.

Negative Correlation

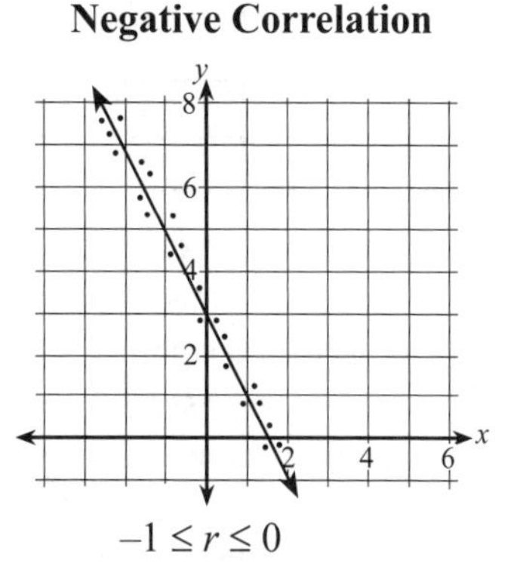

$-1 \le r \le 0$

Residual: It is the difference between the observed y value (y_o) and the predicted y value (y_p) at each observed x value of (x_o). The residuals are another way to examine the correlation of the line of best fit. Ideally the residuals should be as small as possible which would indicate that the regression line was a good fit for the data. The strength of the relationship between the data and the regression function is determined by examining the location of each plotted point compared with the line or curve of best fit.

Residual Plot: It is a scatter plot of the points representing the differences between the observed y_o, and the predicted y_p, at each observed value of x_o. The plotted ordered pairs are (x_o, y_r). The *residual* (y_r) is the difference between an observed "y" value (y_o) and the "y" value (y_p) predicted by the equation for the line of best fit, also called the regression line. A good fit would be indicated if most of the points of the residual plot are near the line $y = 0$. The residuals can be used to discuss the correlation of the line of best fit.

Example Determine whether the line of best fit shown in the bird feeder example on page 164 is a good fit for the data. This is called the correlation.

Plan: Use the regression function found using the calculator for best accuracy. Substitute the observed values of x, (x_o). in the function to find the corresponding predicted value of y_p. Find the difference between the predicted value of y_p, and the observed y_o value. We will call the difference, y_r. The calculation of y_r is: $y_o - y_p = y_r$.

Bird Feeders (x_o)	Bird Seed (lbs) (y_o)	$y_p = 3x_o - 0.8$	$y_o - y_p = y_r$	Residual Plot (x_o, y_r)
1	3	2.2	0.8	(1, 0.8)
2	4	5.2	–1.2	(2, –1.2)
3	9	8.2	0.8	(3, 0.8)
4	10	11.2	–1.2	(4, –1.2)
5	15	14.2	0.8	(5, 0.8)

Analysis: After plotting a graph of the residuals, we can see that most of the y points are close to $y = 0$. The observed points have small differences or residuals with respect to the line of best fit. This indicates that the line of best fit is a good fit. The data and the line of best fit have a strong correlation.

Residual Plot

Causation: Even a strong positive or negative correlation does not necessarily imply cause and effect. In the bird feeder example, the number of pounds of bird seed used does appear to be caused by how many bird feeders a person has. In other cases a strong association could be caused by other variables. Concluding that "*x causes y*" cannot be proved simply with the correlation coefficients and residuals.

Conclusion: There are many other quantitative statistical calculations that can be performed using data and their associated graphing calculator functions which will be studied in later math courses. Figure 1, on the previous page, is used to demonstrate a positive correlation. It represents an exponential regression which has a curved "line" of best fit. Other types of regression include logarithmic and power regressions, which are also both curved.

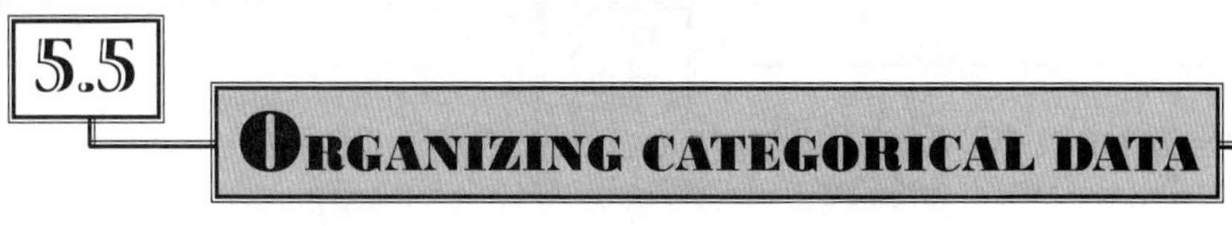

5.5 ORGANIZING CATEGORICAL DATA

TWO WAY FREQUENCY TABLES

Categorical data can be presented using a ***two-way frequency table***. In a two-way table, there are ***two categorical variables*** involved. Within each variable there can be two or more categories included. The actual count of the data is recorded in the two-way table.

Data Set 1 and Figure 1 below show just such a distribution of categorical data. They represent the collection of data about freshman girls and boys who ride or walk to school this year. The two categorical variables are *gender* and *travel method*. The categories for gender are boy or girl. The categories for travel method are ride or walk. This is the third consecutive school year that this survey has been conducted. All freshmen were surveyed each year.

Data Set 1: There are 200 freshmen, 75 boys and 125 girls.
52 boys ride to school and 79 girls ride to school.
The remainder of the freshmen walk.

Example Make a two-way table and record the data. (Give the table all necessary and appropriate titles.)

Figure 1

HOW FRESHMEN GET TO SCHOOL		
Travel Method	**Gender**	
	Boys	**Girls**
Ride	**52**	**79**
Walk	(75 – 52 =) **23**	(125 – 79 =) **46**

The table requires a main title (How Freshmen Get to School). It then requires a label for each *variable* (gender, travel method) and for each *category* of the 2 variables (boy, girl) and (ride, walk).

Gender is called the ***column variable*** because the count or frequency for boys or girls is recorded in the columns. Travel method is called the ***row variable*** because the count or frequency of riders or walkers is recorded in the rows.

The data *recorded* in the body of the table are called ***joint frequencies*** because each cell simultaneously or jointly contains the count of the data items for both the row variable (travel method) and the column variable (gender): boys who ride, boys who walk; girls who ride, and girls who walk.

Note: For demonstration purposes, the calculations for how many of the boys and of the girls walk are shown in Figure 1. It is not necessary to show calculations on the table.

Data Set 2: At noon, on a Monday, data set 2 is collected from three parks. At the Washington Park playground, 15 children are playing on the slide, 12 are on the merry-go-round, 4 are on the see-saws, and 3 are on the swings. At Lincoln Park, 10 are on the slide, 8 are on the merry-go-round, none are on the see-saw, and 2 are on the swings. At the playground at Roosevelt Park, there are 5 on the slide, 10 on the merry-go-round, 6 on the see-saws, and 5 are swinging.

Example Make a two-way table and record the data for Data Set 2, with appropriate titles.

Plan: To make a two-way frequency table, first determine what the row and column variables are and what the categories are.

Row Variable: Equipment

Categories: slide, merry-go-round, see-saw, swings.

Column Variable: Park

Categories: Washington, Lincoln, Roosevelt

Figure 2

Number of Children Using Playground Equipment at Noon			
Equipment	**Name of Park**		
	Washington	**Lincoln**	**Roosevelt**
Slide	**15**	**10**	**5**
Merry-go-round	**12**	**8**	**10**
See-saw	**4**	**0**	**6**
Swings	**3**	**2**	**5**

It is important to note that in this example each of the 2 variables (name of park, playground equipment) has more than two categories. But the table itself is still representing two variables and is still therefore called a two-way frequency table. It will have more than two rows and/or columns of recorded data.

Marginal Frequencies

To our table, an additional column and row called "Total" are added. The extra column and extra row are added for calculating the sums of each column and row. These sums are called the ***marginal frequencies*** – they are located in the margins of the chart! Marginal frequencies are sometimes called the marginal distribution.

Note: If the marginal frequencies are not on the table, ***always*** calculate them before doing any further interpretation of the data.

Calculate the content of each cell in the margins by adding the frequencies to the left of the cell, or above the cell. Each cell in the margins contains the marginal frequency for that row or column. Again, the calculations are shown for demonstration.

Example Create a two-way table with marginal frequencies for Data Set 1. (***Hint:*** Refer to Figure 1 on page 169.)

Figure 3

How Freshman Get to School			
Travel Method	**Gender**		**Total** (marginal row frequency)
	Boys	**Girls**	
Ride	52	79	(52 + 79 =) **131**
Walk	23	46	(23 + 46 =) **69**
Total (marginal column frequency)	(52 + 23 =) **75**	(79 + 46 =) **125**	(131 + 69 =) **200** & (75 + 125 =) **200**

The *marginal frequencies* for the *column variable*, Gender, are 75 for boys and 125 for girls. There are 75 boys in the data and 125 girls. The *marginal frequencies* for the *row variable*, travel method, are 131 for ride and 69 for walk. 131 boys and girls ride and 69 boys and girls walk. The overall total, the number of people from whom data were collected is in the bottom right corner of the table, 200 people. Check your work. It must be the same sum for the frequencies to its left as well as for the frequencies above it.

What do the data show? A quick interpretation of the frequencies in Figure 1 shows that more girls than boys ride to school. But notice that there are many more girls than boys who were surveyed. The analysis of the data can be misleading. Other calculations using percents are performed to help make it clearer. (See also Relative Frequencies on page 172)

Example Create a marginal distribution for Data Set 2. (***Hint:*** Refer to Figure 2 on page 170.)

Figure 4

Number of Children Using Playground Equipment at Noon				
Equipment	**Name of Park**			
	Washington	**Lincoln**	**Roosevelt**	**Total**
Slide	**15**	**10**	**5**	**30**
Merry-go-round	**12**	**8**	**10**	**30**
See-saw	**4**	**0**	**6**	**10**
Swings	**3**	**2**	**5**	**10**
Total	**34**	**20**	**26**	**80**

M*arginal frequencies for the column variable*, Park, are 34, 20, and 26. There were 34 children at Washington, 20 at Lincoln, and 26 at Roosevelt. *Marginal frequencies for the row variable*, Equipment, are 30, 30, 10 and 10. Thirty used the slide, 30 used the merry-go-round, 10 used the see-saw, and 10 used the swings. The total number of children using the playground equipment is 80 at bottom right corner of table.

What do the data mean? It appears from the marginal row frequencies that the slide and the merry-go-round are equally popular with the children. The see-saw and swings each have only one third the frequency of the slide or merry-go-round. This might suggest that children don't like the see-saw and swings. However, it should be noted that a see-saw only holds two people, so the number using it at one time is physically limited.

- In some data collections a "lurking variable" is involved, when unknown information about one variable can impact the interpretation of the data. The see-saw only holding two people is a possible lurking variable.

RELATIVE FREQUENCY AND TWO-WAY RELATIVE FREQUENCY TABLES

Relative frequency is an additional calculation using the two-way frequency table that enables more complete analysis of the data. It is a percent or decimal calculated by taking a number of recorded data and dividing it by a total. Three types of relative frequency are ***joint, marginal,*** and ***conditional.***

In each situation it is necessary to decide what fraction is needed to get the percent desired. First determine which group represents the ***total amount you want a percent of***. The count of that total group is the denominator of the fraction used to find the relative frequency. The numerator of the fraction is the count of the specific part of that group which meets the required characteristics.

Relative frequency can be expressed as a fraction, decimal, or percent. The relative frequencies calculated should have a sum of 1.0 or 100%, although sometimes, due to rounding, there are slight discrepancies.

For demonstration purposes, the calculations used are shown on the following tables. They do not need to be written on the table, only the quotient in decimal or percent form needs to be written. The cells not needed for that particular type of relative frequency are erased. All the following tables are based on (Data Set 1 on page 169.)

Joint Relative Frequency: Using the information from a two-way frequency table allows us to develop a two-way relative frequency table. The data in a relative frequency table contain the relative frequency of each of the entries in the two-way frequency table. The relative frequency is expressed as a percent or a decimal.

For easy reference, here is a completed, summarized copy of the two-way frequency table of Data Set 1. Remember that these are the "recorded" values - the actual count of the people in the group who meet the criteria of both the row variable and the column variable. Each row and column is totaled.

Figure 5

How Freshmen Get to School			
Travel Method	**Gender**		
	Boys	**Girls**	**Total**
Ride	52	79	**131**
Walk	23	46	**69**
Total	**75**	**125**	**200**

Joint relative frequency determines how the frequency of one cell (numerator) of joint recorded data in the two-way frequency compares with the total count of items (denominator) in the entire data collection.

Example Create a relative frequency table to show the relative joint frequencies for Data Set 1. What conclusions could be drawn from the joint relative frequency calculations? (***Hint:*** Refer to Figure 5.)

Figure 6

Joint Relative Frequency - How Freshmen Get to School			
Travel Method	**Gender**		
	Boys	**Girls**	**Total**
Ride	$\frac{52}{200}$ **= 0.26 or 26%**	$\frac{79}{200}$ **= 0.395 or 39.5%**	
Walk	$\frac{23}{200}$ **= 0.115 or 11.5%**	$\frac{46}{200}$ **= 0.23 or 23%**	
Total			$\frac{200}{200}$ **= 1.0 *or* 100%**

Each *numerator* is the count in the *recorded cell* from the two-way table, and the *denominator* for each calculation is the *total number of people* in the study.

The joint relative frequency indicates what portion of the total freshmen are boys who ride, (26%), boys who walk (11.5%), girls who ride (39.5%), and girls who walk (23%).

Analysis: The joint relative frequencies show that the highest percentage of the freshmen are girls who ride to school. The next highest percentage of the freshmen would be boys who ride to school. Next would be freshman girls who walk and lastly freshman boys who walk.

5.5

Marginal Relative Frequency makes a comparison between the total of each category of each variable and the overall total of the data. For example, we might want to know what percent of the total is in the category "boys" of the variable "Gender", i.e., what portion of the total 200 students is boys. We would use...

$$\frac{\textit{75 total boys}}{\textit{200 total students}} = 0.375 \ \textit{or} \ 37.5\%$$

Example Complete a marginal relative frequency table for Data Set 1. Describe when marginal relative frequency is helpful and describe some conclusions that can be drawn about the data. (***Hint:*** Refer to Figure 5 on page 173.)

Figure 7

MARGINAL RELATIVE FREQUENCY - HOW FRESHMAN GET TO SCHOOL			
Travel Method	**Gender**		
	Boys	**Girls**	**Total**
Ride			$\frac{131}{200}$ = **0.655 or 65.5%**
Walk			$\frac{69}{200}$ = **0.345 or 34.5%**
Total	$\frac{75}{200}$ = **0.375 or 37.5%**	$\frac{125}{200}$ = **0.625 or 62.5%**	$\frac{200}{200}$ = **1.0 *or* 100%**

Each *numerator* is the count in the *recorded cell* from the two-way table, and the denominator for each calculation is the *total number of people* in the study, 200.

When would you need to calculate marginal relative frequency from the table?

Marginal relative frequency provides us with information about what portion of the total 200 freshmen is boys or girls; and what part of the 200 students rides or walks to school. 37.5% are boys, 62.5% are girls; 65.5% ride and 34.5% walk to school.

Analysis: From these calculations we can say that there is a higher percent (62.5%) of *girls* in the *total students* than *boys* (37.5%). We also can say that about 2/3 of the *students* (65.5%) get a *ride* to school, while only about a third (34.5%) *walk.*

5.5

<u>**Conditional Relative Frequency**</u> is a comparison within *one variable* (either row or column) of the data in each *category* it contains compared with the *total in that particular row or column.*

Example *Of the riders* (our condition), is there a greater percentage of girls or boys?

Plan: Remember, the *denominator* of the fraction is the *total amount you want a percentage of.* Therefore, we only need to look at the totals for riders to find the denominator.

$$\frac{52 \text{ boys who ride}}{131 \text{ total}} = 0.40 \text{ or } 40\%$$

$$\frac{79 \text{ girls who ride}}{131 \text{ total}} = 0.60 \text{ or } 60\%$$

Analysis: A greater percentage of the riders are girls.

Example Complete a relative frequency table *for* the gender of riders or walkers. Give a brief analysis of the results.

Plan: This example is asking for a complete table for the relative frequencies for gender (column variable) *of* riders or walkers (row variable). The numerators for these calculations are the counts in the recorded cell under the column required. The denominator is the total for that row.
(***Hint:*** Refer to Figure 5 on page 173.)

Figure 8

CONDITIONAL RELATIVE TRAVEL FREQUENCIES - HOW FRESHMEN GET TO SCHOOL			
Travel Method	**Gender**		
	Boys	**Girls**	**Total**
Ride	$\frac{52}{131} \approx$ **0.40 or 40%**	$\frac{79}{131} \approx$ **0.60 or 60%**	**131**
Walk	$\frac{23}{69} \approx$ **0.333 or 33.3%**	$\frac{46}{69} \approx$ **0.667 or 66.7%**	**69**
Total			

Analysis: From these relative gender frequencies we find that about 40% of the 131 students who ride to school are boys and 60% are girls. Of the 69 students who walk, 33.3 % are boys and 66.7 % are girls.

Example Suppose we wanted to know that *of the boys*, (our condition) how do a greater percentage get to school?

Plan: Remember, once more, that the *denominator* of the fraction is *the total amount you want a percentage of.* Therefore, we need to only look at the total for boys to get that denominator.

$$\frac{52 \text{ boys who ride}}{75 \text{ total boys}} = 0.693 \text{ or } 69.3\%$$

$$\frac{23 \text{ boys who walk}}{75 \text{ total boys}} = 0.307 \text{ or } 30.7\%$$

Analysis: A greater percentage of boys ride to school than walk.

Example Complete and discuss a conditional relative frequency chart that demonstrates what percent of the boys ride and what percent of the boys walk to school. Display the same information for the girls.

Plan: Make a table for conditional relative frequencies for travel method (the row variable) of gender (column variable).

(***Hint:*** Refer to Figure 5 on page 173.)

Figure 9

CONDITIONAL RELATIVE GENDER FREQUENCY – HOW FRESHMEN GET TO SCHOOL			
Travel Method	**Gender**		
	Boys	**Girls**	**Total**
Ride	$\frac{52}{75} \approx$ **0.693 or 69.3%**	$\frac{79}{125} \approx$ **0.0.632 or 63.2%**	
Walk	$\frac{23}{75} \approx$ **0.307 or 30.7%**	$\frac{46}{125} \approx$ **0.368 or 36.8%**	
Total	**75**	**125**	

The numerators here are the counts in the appropriate recorded cell, and the denominators are the count in the total for that column of gender categories. These conditional relative frequency values tell us that of the 75 boy students, 69.3% ride and 30.7% walk. Out of the 125 girl students, 63.2% ride and 36.8% walk.

Analysis: From this we can say that a higher percentage of the boys ride to school than the percent of girls who ride to school.

Note: Examining the relative frequencies gives a different viewpoint of the information than just using the two-way frequency table. Looking back at the two-way frequency table for Data Set 1, we saw that more girls than boys ride to school. However, in this last table, conditional relative frequency with travel methods, we can determine that a higher percentage of boys ride than the percentage of girls.

Final Analysis: Since this is the third consecutive year that all the freshmen in this high school have been surveyed, we can compare the results of this year's survey with the two from the previous two years. Let's look at the trend in whether students walk or ride (the row variable). The marginal relative frequencies seem like a reasonable place to compare.

Marginal Relative Frequencies 2 years ago showed that 60% of the freshmen rode to school and 40% walked. In the data from last year, 63.2% students rode and 36.8% walked. This year 65.5.% ride and 34.5% walk.

It appears from this comparison of data that there is a trend for more students to ride to school than walk. Since the percent of riders has increased over the last 3 years, it seems likely that next year that percentage would increase again. In discussing this, we can also say there is a trend for a lower percentage of the students to walk to school each year.

Correlations to Common Core State Standards

Common Core State Standards	Unit # . Section #

The Real Number System (N-RN)

N-RN.3 .... 2.1, 2.2

Quantities (N-Q)

N-Q.1 .... 2.3
N-Q.2 .... 2:3
N-Q.3 .... 2.4, 2.5

Seeing Structure in Expressions (A-SSE)

A-SSE.1 .... 2.1, 3.1, 3.2, 3.3
A-SSE.2 .... 3.4
A-SSE.3 .... 3.4

Arithmetic with Polynomials and Rational Expressions (A-APR)

A-APR.1 .... 3.5
A-APR.3 .... 3.4

Creating Equations (A-CED)

A-CED.1 .... 3.6, 3.7, 3.10
A-CED.2 .... 3.7, 3.8
A-CED.3 .... 3.11, 3.12
A-CED.4 .... 3.3, 3.13

Reasoning with Equations and Inequalities (A-REI)

A-REI.1 .... 3.6, 3.7
A-REI.3 .... 3.6, 3.8, 3.10
A-REI.4 .... 3.13
A-REI.5 .... 3.9
A-REI.6 .... 3.9
A-REI.10 .... 3.8
A-REI.11 .... 3.9
A-REI.12 .... 3.10, 3.11, 3.12

Correlations to Common Core State Standards

Common Core State Standards	Unit # . Section #
Interpreting Functions (F-IF)	
F-IF.1	4.1, 4.2, 4.8
F-IF.2	4.1
F-IF.3	4.3, 4.8, 4.10
F-IF.4	4.2, 4.3, 4.4, 4.6
F-IF.5	4.1
F-IF.6	4.6
F-IF.7	3.8, 3.9, 4.3, 4.4, 4.6
F-IF.8	3.13, 4.6
F-IF.9	4.3, 4.6, 4.9, 4.10
Building Functions (F-BF)	
F-BF.1	4.7, 4.8, 4.9, 4.11
F-BF.3	4.5, 4.7, 4.8
Linear, Quadratic and Exponential Models (F-LE)	
F-LE.1	4.8, 4.9
F-LE.3	4.6, 4.9
F-LE.2	4.6, 4.8, 4.9
F-LE.5	3.13, 4.6, 4.8
Interpreting Categorical and Quantitative Data (S-ID)	
S-ID.1	5.3
S-ID.6	5.4
S-ID.2	5.2, 5.3
S-ID.7	4.5
S-ID.3	5.2, 5.3
S-ID.8	5.4
S-ID.5	5.1, 5.5
S-ID.9	5.1, 5.4

INDEX

Q

R

S

Index